MARION P. HARTHILL
IRENE O'CONNOR

Common Mosses of the Pacific Coast

NATUREGRAPH PUBLISHERS

Library of Congress Cataloging in Publication Data CIP

Harthill, Marion Paul, 1909–

Common mosses of the Pacific coast.

1. Mosses–Pacific States–Identification. I. Title.

QK541.5.P3H37 588'.2'0979 74-31294

ISBN 0-87961-025-5 Cloth Edition
ISBN 0-87961-024-7 Paper Edition

NATUREGRAPH PRESS

Naturegraph Publishers Inc., Healdsburg, California 95448

CONTENTS

PREFACE

The author hopes that this guide will open a new field of interest for visitors to the Pacific Coast. Few realize that more than 300 different kinds of mosses festoon the giant trees of this area, and cover their precipitous hillsides. By cushioning the fall of the rain and slowing its rush to the Pacific, they reduce erosion, and the restful green of the hills is preserved for posterity.

This study was made possible by financial aid from The National Parks Association in the summer of 1960, and from the American Academy of Arts and Sciences during the summer of 1961.

Bryologists all over the world have aided the author in many ways, but he particularly wishes to acknowledge the continual and generous help that he has received from Dr. H. S. Conard, author of *How to Know the Mosses and Liverworts.* Dr. Conard collected in the Olympics at the turn of the century and taught many years at Grinnell College in Iowa. Until his death, at the age of ninety-seven, he resided at Lake Hamilton, Florida.

Marion P. Harthill

To him who in the love of nature holds communion with her visible forms she speaks a various language.

Bryant

INTRODUCTION

The Pacific Coast has been described by Dr. H. S. Conard as a bryologist's paradise. Although much of the virgin forest, through which he roamed as a young botanist in 1902, is no more, remnants have been preserved as National Parks and Wilderness Areas. The role that mosses play in preventing soil erosion is an important one. Holding many times their own weight in water, mosses retard the downward rush of the water, and the soil stays on the hills.

In our imagination, let us sweep back through the ages to a time when the Pacific Northwest was emerging from the grip of the ice sheets. Melting glaciers left a desert of bare rock, moraines, and gravelly outwash plains. No plant life was visible, but, borne on the wings of the swirling winds, tiny lichen soredia settled to earth and formed the first vegetal covering of the new land. Lichens are unique in their ability to sustain themselves on bare rock, secreting organic acids that slowly crumble the surface to which they cling so tenaciously.

The air bears many bits of living protoplasm, protected by tough coats, awaiting the proper time and place to burst into life and clothe the earth with green. Microscopic in size, moss spores settled to earth. Finding a bit of soil and moisture under the lichens, they grew and divided until they produced the moss protonemata or 'first threads' that formed a thin mat over the substrate. From these protonemata grew the moss plants that in turn produced more spores.

As the mosses spread, the seeds of the higher plants found a satisfactory seed bed, germinated, and the forests were under way. Mosses do not always require lichen pioneers. The author has seen *Grimmia alpestris* growing on as bare a rock as ever did a lichen. Wind-blown dust settles in the cracks formed by frost, and is immediately colonized by mosses. Watch a new highway cut for a few years and notice the mosses slowly take over, on either bare rock or sterile gravel.

Lying across the path of the westerly winds sweeping in off the Pacific Ocean, the mountains of the Pacific Coast are annually drenched with up to ten or twelve feet of water on their western slopes. Fortunately, little of this falls during the tourist season, so that the heavy rains of winter create the moss-upholstered rain forest for our summer pleasure. The heavy precipitation on the rain forest gives rise to the most luxuriant growth of mosses outside the tropics.

The trunks of fallen trees become moss-covered mounds. Boulders, house size, are completely enshrouded in bryophytic verdure. The trunks and larger limbs of trees are completely encased with green padding. Lace-like, the pendant mosses hang from the smaller limbs in curtains. A greenish luminosity pervades the whole, leaving one with an 'Alice in Wonderland' feeling.

To many people, any small, green plant is a 'moss'. 'Moss Campion', *Silene acaulis,* grows in the Arctic-Alpine zone. A perfect clump of 'moss', it is in reality a vascular plant that has tiny pink blossoms in season and is a member of the Carnation Family. 'Reindeer Moss' is a lichen, as is the Old Man's Beard that hangs abundantly from limbs throughout the forest. How, then, do we recognize a moss? Mosses have extremely thin leaves, only one cell thick in many cases. The leaf may or may not have a midrib or costa. It never has veins, as does a vascular leaf. Mosses do not have roots, as do flowering plants. Thin, hair-like rhizoids serve the same purpose, and anchor the plant to the earth. Moss stems do not have vessels for conducting food and water, as do the higher (vascular) plants. They have chlorophyll, and manufacture their own food in place. Water is absorbed from the outside. Mosses are extremely tenacious of life. A dried Grimmia that had been in a herbarium for a year, revived and put forth new growth when it was given water and sunlight. Pour a few drops of water on a completely dry moss that is baking under the intense sunlight above timberline and notice how quickly it revives. Some species can be ground to powder between your fingers and when moistened on a favorable substrate, will put forth new growth and recreate themselves from the cell fragments.

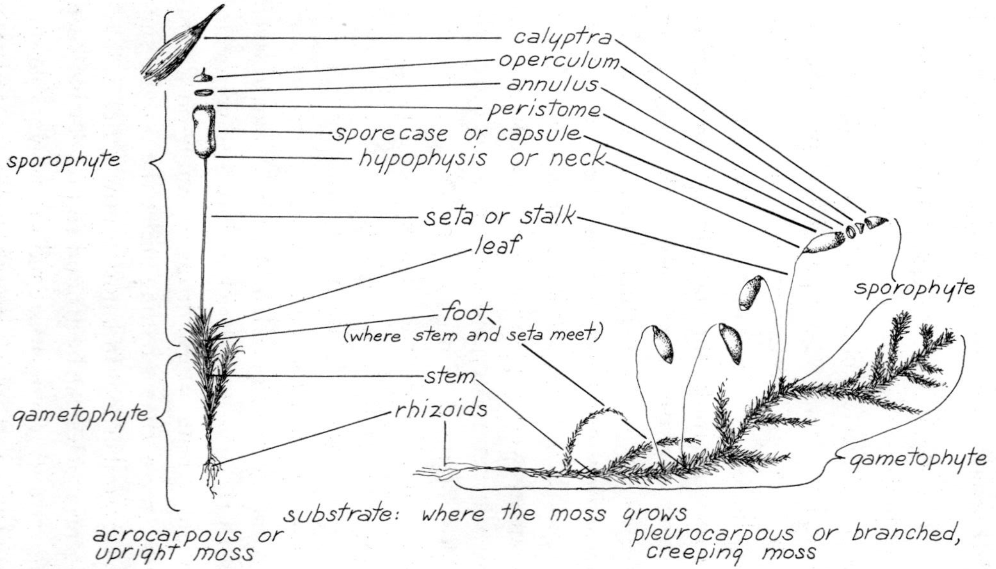

Anatomy of Acrocarpous and Pleurocarpous Mosses.

Liverworts or hepatics may cause you some confusion. None of them resembles the mosses described in this key. The moss-like ones tend to have rounded leaves with no midrib. Spore capsules are short-lived and are borne on watery setae. The leaves in mosses do not appear scale-like as they do in hepatics. After a little experience, you will wonder why you were ever confused.

Mosses are divided into two general groups. The upright, or acrocarpous, mosses have setae and capsules arising from the *tip* of the stem. This may not seem to be the case in some species like the Grimmias. Here the side shoots may grow faster and become longer than the original stem, but the setae still grow from the tip of the original stem.

In the other group, the pleurocarpous mosses, the plants creep over the substrate and branch freely, the spore capsules arising *along* the stem. These two groups must be readily distinguished, as this is the first separation that the key makes.

The growth and reproduction of moss plants make an interesting story. As stated earlier, the moss spores are air borne and may be worldwide in their distribution. Settling down on a favorable spot, they absorb water and begin their growth. Thin mats of protonema spread over the substrate in which their growth starts. The thread-like protonema resembles the green algae from which mosses may have evolved. Soon, stems arise with leaves, and we recognize the typical moss plant. The botanist calls this plant the gametophyte. These are the plants that interest us. This gametophyte does not produce spores. It produces gametes, or eggs and sperm.

Some of the Mniums have the male 'flowers' on one plant and the female on another. The female 'flowers', or archegonia, are jug-shaped and produce one egg each. More conspicuous than the archegonia are the antheridia, or male 'flowers', which are cigar-shaped and tightly clustered in a flower-like head. At the proper time each antheridium produces myriads of tiny swimming sperms that propel themselves by means of tails (flagella) through the surface moisture on the leaves to enter the mouth of the archegonium and fertilize the egg, which then grows and forms the spore capsule.

In the Badge Mniums, *Mnium insigne,* one plant will appear to have a flower on the tip of the stem. Microscopic analysis will reveal many tightly packed cigar-shaped antheridia. Each of these antheridia will produce many sperm cells, tiny male germ cells that are able to swim by themselves.

On another plant, the tip will have a jug-shaped archegonium nestled among the leaves. This archegonium will have one cell in the bottom, the egg. When the antheridia are mature, and after a rain or heavy dew, swarms of sperm are released. Some of these sperm find their way to receptive archegonia and, attracted by a sweetish secretion from the egg, they swim down the neck. The first sperm to reach the ovum enters it, and fertilization is effected. An interesting corollary to this is the fact that the same sweetish secretion from the egg attracts the sperm cell in the higher mammals such as man.

After the sperm has entered the egg, development begins, and a spear-like growth emerges from the archegonium and reaches skyward. In emerging from the jug-shaped archegonium, it pushes off the top and lifts it: the calyptra. This sharp-pointed stalk becomes the spore-producing plant or sporophyte. We commonly call it the spore capsule. As it matures, we find the capsule at the top of the seta. The top of the capsule has a lid, or operculum, which in its early life is covered with the remains of the archegonium, the calyptra. The calyptra is shed first and then the lid or operculum. This exposes the peristome, a ring of teeth around the mouth of the capsule. *Peri* means around, as in the word perimeter; *stome* means mouth. These teeth, although they vary in number, with the species, from four to sixty-four, are always found in a multiple of four. King George's Moss, *Tetraphis pellucida,* has four, and the Juniper Haircap, *Polytrichum juniperinum,* has sixty-four. Many mosses have single peristomes of sixteen teeth. In others, the sixteen are split to make it appear that there are thirty-two. In a large number of species there is an inner and an outer peristome. The inner teeth are transparent and have hair-like cilia between them. All of these things are important characteristics in the scientific classification of mosses.

The function of the peristome teeth is to close

the mouth of the capsule in damp weather. When the sun shines again, they open to permit spores to escape. This hygroscopic characteristic of the peristome is sometimes uncanny to observe through a microscope. As you look at some capsules, the teeth are rhythmically opening and closing. There seems to be a close correlation between this movement and yourself. Soon you discover that, as you exhale, your moisture-laden breath dampens the sensitive teeth and they bend in to close and protect the spores. As you inhale, dry air is pulled in and the teeth dry and open, to allow the spores to escape. What a sensitive hygrometer one of these teeth would make!

I have not attempted to describe the variations that occur in the reproduction of different mosses. Some plants have the archegonia and antheridia on the same plant. Sometimes they occur in the same cluster on the tip of the stem, the archegonium in the center with the antheridia, mixed with hair-like paraphyses, clustered around it. This is the case in the Twisted Moss, *Tortula princeps.* Some mosses seldom fruit, merely spreading by vegetative growth of the gametophyte generation. A few, like the Lover's Moss, *Aulacomnium androgynum,* bear little buds or gemmae, on the tip of a stem, that grow directly into new moss plants. Ocean Ulota, *Ulota phyllantha,* and Hartman's Grimmia, *Grimmia hartmanii anomala,* have the gemmae on the leaf tips. Others, as I have stated before, are able to grow from mere fragments of the plant, much as you can start an African violet from a portion of the leaf.

From the standpoint of actually being harvested and used by man, mosses have little or no economic value. Semidecayed Peat Moss, as it is found in bogs, is well known as a soil conditioner and mulch. Because of its tremendous absorptive capacity, Peat Moss was used in World War I to pack the wounds of soldiers.

The large, pleurocarpous mosses, such as the Little Shaggy, *Rhytidiadelphus loreus,* are very popular among gardeners for use in hanging baskets. These uses are all very trivial; but beyond dollar evaluation is the role played by mosses in hillside water retention.

When one drinks from the crystal clear waters of Barnes Creek on Lake Crescent (Olympic National Park), he wonders how it could ever have carried enough debris to build up the delta we now call Barnes Point. Dr. J. L. Erspamer, author of *Forests and Trees of the Olympic National Park,* suggests that the bulk of the delta was laid down soon after the glaciers melted. At this time, the running waters had a field day. With no plant cover to stay their force, every rainstorm would cause a flash flood. Our clear stream of today was a roaring torrent of thick brown mud, gravel and rock, pouring over the precipitous edge of Lake Crescent to fill in that portion of the lake and form a delta. Today, all is different. The hillsides have a plant cover that laces the top soil together. Surface water is absorbed by the moss, and eroding torrents have become the beautiful creeks that tumble down the mountain canyons.

Boulders, the size of house trailers, lie on the valley floor. Picture their headlong descent from the pinnacles of Mount Storm King when they were loosened by the uninhibited erosion of yesteryear. Now they lie placidly in the shade of giant Douglas-firs covered with moss and ferns.

If eyes were made for seeing—then beauty is its own excuse for being. Agreeing with the poet, we realize that, last but not least, the aesthetic value of mosses is no mean thing. Compare the almost mossless forests of the arid interior with heavily cushioned woodlands of the coast, the hot glare of the sun is relieved with a restful green—and one relaxes. And I guess that is why one comes to the Pacific coast of the continent for one's vacation.

KEY TO COMMON MOSSES

A KEY TO THE COMMON MOSSES OF THE PACIFIC COAST

BEFORE USING THIS KEY

A botanical key is an either/or sort of thing. Your specimen will be either an erect moss with a spore capsule at its tip, in Part A of the key, or a creeping plant with spore capsules along its stem, in Part B of the key.

In either case it will have been found in a habitat that is constantly wet or generally dry. It will be large or small. Normally a key has only two choices. In this key it has been expedient to have as many as three choices, but this is exceptional.

The author feels that this study should cover the common, conspicuous mosses that the visitor is likely to notice. He will be happy if the user realizes that the key is intended only to spark his interest in the Bryophytes. If he is irked by the exclusion of a large number of mosses that are not described, then he is ready for Dr. H. S. Conard's *How to Know the Mosses and Liverworts,* published by the Wm. C. Brown Co. of Dubuque, Iowa. Dr. Elva Lawton of the University of Washington has recently published a manual describing, in great detail, all the mosses of the Pacific Northwest.

Comments on and criticisms of this key will be welcomed by the author from all interested students.

(**Note:** A pocket magnifier of 6x to 10x is very useful in identifying mosses with this key.)

I. LOWLAND MOSSES, sea level to 3000 feet. (See also p. 29.)

A Moss plants upright; capsules if present, on the tips of the stems (acrocarpous). (For B, see p. 23.)

1a Habitats continuously wet: either in water or splashed with water for the greater part of the year.

2a Plants large, 2 to 3 inches and larger.

3a Soft, light-green mosses of swamps and rain forests that are continuously wet.

PEAT MOSS
Sphagnum species

3b Harsh, blackish mosses, with globular spore capsules, growing on rocks, submerged, or partly submerged, at the edges of the larger streams.

SCOULER'S MOSS
Scouleria aquatica Hook.

2b Plants smaller, under 2 inches; habitat similar to that of 3b; dark green or brownish.

4a Much branched, brownish mosses with urn-shaped capsules that have no visible setae (stems).

COMMON GRIMMIA
Grimmia alpicola Hedw.

4b Less branching, greener, with cylindrical spore capsules that have very long peristome teeth; also common on rocks in drier habitats.

GREEN FRAYED-CAP MOSS
Rhacomitrium varium (Mitt.) Lesq. & James

1b Habitats not continuously wet; mainly dependent on current weather.

5a Coarse mosses of drier habitats.

6a Upright stems, with leaves radiating in all directions.

7a Leaves pointed, with plane edges; calyptra (cap), over spore capsule, hairy. If the calyptra has been shed, look for capsules that seem to have a skin stretched across the top.

8a Capsule ridged lengthwise; leaves stiff, with reddish tips.

JUNIPER HAIRCAP
Polytrichum juniperinum Hedw.

8b Capsule cylindrical; leaves soft at all times.

BEARDED MOSS
Pogonatum alpinum (Hedw.) Roehl.

7b Leaves pointed, but with very wavy edges; no hair on calyptra.

QUEEN CATHERINE'S MOSS
Atrichum undulatum (Hedw.) P. Beauv.

6b Shorter, much branched mosses covering the surfaces of rocks and dry gravelly soil; grayish appearance contributed by transparent leaf tips.

GRAY FRAYED-CAP MOSS
Rhacomitrium canescens (Hedw.) Brid.

5b More delicate plants.

9a Plants tree-like.

10 Mosses like tiny palm trees 'in a wind storm'; prefer moist, shady places.

MENZIES'S TREE MOSS
Leucolepis menziesii (Hook.) Steere ex L. Koch

9b Plants not tree-like.

11a Plants with curled leaves or setae.

12a Plants growing on wood or litter.

13a Plants in clumps or mats.

14a Large clumps of curly leaves; spore capsules bent, with a long sharp calyptra.

CURLY HERON'S-BILL MOSS
Dicranum fuscescens Turn.

14b Thinner clumps or mats with erect yellow capsules.

WEIS DICRANUM
Dicranoweisia cirrhata (Hedw.) Lindb. or Milde

13b Plants in 'buttons' on branches of broadleafed trees; calyptra hairy.

15a Leaves curled, but normal otherwise; spore capsules stubby.

OBTUSE ULOTA
Ulota obtusiuscula C. Mull. & Kindb.

15b Leaves curled, but with tiny rod-shaped gemmae on tips; sea coast only, in reach of salt spray.

OCEAN ULOTA
Ulota phyllantha Brid.

12b Plants growing on rock or soil.

16a Plants growing in clumps.

17a Small clumps of stiff, curly leaves growing in reach of salt spray on sea cliffs.

OCEAN GRIMMIA
Grimmia maritima Turn.

17b Large clumps of soft, curly leaves; spore capsules apple-shaped; grows in moist shady places; common.

APPLE MOSS
Bartramia pomiformis Hedw.
var. *elongata* Turn.

16b Plants not growing in clumps.

18a Plants on rock; small plants; leaves tightly curled, with erect yellow spore capsules.

WEIS DICRANUM
Dicranoweisia crispula (Hedw) Lindb. ex Milde

18b Plants on soil; shows preference for old campfire sites; long, wavy setae with oddly shaped spore capsules; leaves folded closely.

TWISTED-CORD MOSS
Funaria hygrometrica Hedw.

11b Plant leaves and setae not curled.

19a Plants growing on wood.

20a Generally found on live trees or sound wood.

21a Large clumps with leaves bent all one way; spore capsule bent, with long, sharp calyptra.

BROOM MOSS
Dicranum scoparium Hedw.

21b Long, strap-shaped leaves with stubby capsules; more likely to be found on the broad-leaved trees of the rain forest.

STRAIGHT-HAIRED MOSS
Orthotrichum species

20b Plants growing on decaying logs, old stumps, etc.

22a Straight, yellowish-green leaves, with erect spore capsules; tips of leaves often broken off.

BROKEN-LEAF MOSS
Dicranum tauricum Sapeh.

22b Small, globular fruiting bodies (not spore capsules) on slender stalks.

LOVER'S MOSS
Aulacomnium androgynum (Hedw.) Schwaegr.

19b Plants growing on rock or soil.

23a Plants not growing on rocks.

24a Plants in moist rich soil.

25a Plants sword-fern-shaped when moist; contorted and knotty when dry.

BADGE MNIUM — Sterile form
Mnium insigne Mitt.

25b Plants not fern-shaped; texture of leaves lettuce-like; may have several spore capsules on top of stem or clustered 'flowers' of antheridia.

BADGE MNIUM — Fertile form
Mnium insigne Mitt.

24b Plants not showing preference for rich soil.

26a Plants in rosettes of oval-shaped or long, narrow leaves.

27a Capsules barrel-shaped.

28a Leaves almost round, with

smooth edges.

EARLY MNIUM
Mnium punctatum

28b Leaves pointe[d] with toothed e[d]ges; mouth [of] capsule red.

RED-MOUTHED MNIUM
Mnium spinulosum B.S.G.

27b Capsules pear-shaped.

29a Oval-shaped le[a]ves.

BRYUM
Bryum species

29b Long, narr[ow] leaves.

NODDING POHLIA
Pohlia nutans (Hedw.) Lindb.

27c Purple, cylindrical ca[p]sules, deeply furrow[ed] lengthwise.

PURPLE HORNTOOTH
Ceratodon purpureus (Hedw.) Brid.

23b Plants growing on rocks.

30a Large, dense clumps, with tang[led] masses of red hair at the bases [of] stems.

RED-HAIRED MOSS
Anacolia menziesii (Turn.) Par.

30b Coarse mosses with long setae; slightly curved capsules, with long, twisted, tubular peristomes; teeth separate near ends.

TWISTED MOSS
Tortula princeps De Not.

26b Plants not in rosettes.

31 Skirt-like calyptra hanging over capsule.

EXTINGUISHER MOSS
Encalypta rhaptocarpa Schwaegr.

B Moss plants creeping; spore capsules, if present, along stem (pleurocarpous).

1a Habitat continuously wet: either in water or splashed with water for the greater part of the year.

2a Mosses in fast water.

3a Long, trailing submerged mosses; leaves folded down midrib.

4a Somewhat delicate.

NEW MEXICAN FOUNTAIN MOSS
Fontinalis neomexicana Sull. & Lesq.

4b Large, very coarse moss.

GIANT FOUNTAIN MOSS
Fontinalis antipyretica Hedw. var. *gigantea* (Sull.) Sull.

3b Thick beds of light-green moss.

5a Large leaves; plants thick and cushiony; tips of branches sharp and shiny.

GOLDEN SHORT-CAPSULED MOSS
Brachythecium asperrimum (Mitt.) Sull.

5b Small leaves; plants like fine ferns.

SLENDER BEAKED MOSS
Eurhynchium praelongum (Hedw.) B.S.G. var. *stoke* (Turn.) Hook.

5c Thick-stemmed, fleshy mosses; leaves gathered close to t stem; plants clustered on rocks.

BLUNT-LEAVED MOSS
Scleropodium obtusifolium (Hook. or Drumm.) Kinc ex Mac. & Kindb.

2b Mosses in swamps or ponds.

6a Plants tree-like.

TREE MOSS
Climacium dendroides (Hed Web. & Mohr

6b Plants not tree-like.

7a Slender plants with long, curved, pointed leaves; abund paraphyllia (small green growths of various shapes) tween the leaves.

HOOKED MOSS
Cratoneuron species

7b Similar plants, but paraphyllia absent.

SICKLE MOSS
Drepanocladus species

1b Habitats not continuously wet; mainly dependent on current weat

8a Mosses generally found on wood.

9a Mosses found on living trees.

10a Leaves flattened along stem, pale green.

11a Mosses found on trunks and limbs; leaves wi 'permanent wave'.

12a Seta shorter than the capsule.

MENZIES'S NECKERA
Neckera menziesii Hook.
ex. Drumm.

12b Seta longer than the capsule.

DOUGLAS'S NECKERA
Neckera douglasii Hook.

11b Mosses found on down logs, roots, etc.; also with 'permanent wave'.

ELEGANT MOSS
Isopterygium borrerianum
(C. Mull.) Lindb.

10b Leaves not flattened.

13a Plants tree-like; scroll-shaped when dry, like the scroll of a violin.

BALSAM-TREE MOSS
Dendroalsia abietina (Hook.)
Britt.

13b Plants not tree-like.

14a Leaves very tiny, plants large.

15a Long tresses of extremely fine moss.

LACE MOSS
Isothecium spiculiferum
(Mitt.) Ren. & Card.

15b Delicate mats of moss, closely pressed to the bark of tree trunks; red stems; plants seem to "run downhill"; tiny spore capsules with white peristomes.

COILED-LEAF MOSS
Hypnum circinale Hook.

14b Leaves of normal size.

16a Plants feather-shaped.

17a Leaves large, soft, and curly; tips have a drawn-out stringy appearance; individual plants arise from stolons.

CARDOT'S MOSS
Isothecium spiculiferum (Mitt.) Ren. & Card. var. *cardotii* (Kindb. ex Mac. & Kindb.) Crum, Steere & Anderson

17b Leaves more delicate; long, narrow fronds cling closely to smooth bark.

CREEPING BEAUTY
Homalothecium nuttallii (Wils.) Jaeg. & Sauerb.

16b Plants not feather-shaped.

18a Coarse, wire-like moss.

HANGING MOSS
Antitrichia curtipendula (Hedw.) Brid.

18b Coarse clumps or layers of large, golden-green moss.

YELLOW MOSS
Homalothecium fulgescens (Mitt. ex. C. Mull.) Lawt.

9b Mosses found on dead wood and litter of the forest floor.

19a Plants feather-shaped.

20a Large, pointed plant, with very fine leaves; green to golden in color; common.

OREGON BEAKED MOSS
Eurhynchium oreganum (Sull.) Jaeg. & Sauerb.

20b Large, coarse plant, with blunt tip turned back over itself; tip yellowish green, base brown.

GIANT MOSS
Homalothecium megaptilum (Sull.) Robins.

19b Plants not feather-shaped.

21a Mosses fern-shaped.

22a Current year's growth growing out from the top of previous year's growth.

MOUNTAIN FERN MOSS
Hylocomium splendens (Hedw.) B.S.G.

22b Plants in tight curls; spore capsules large.

CURLY HYPNUM
Hypnum subimponens Lesq.

21b Mosses irregular in shape.

23a Plants growing close to the ground; greenish white; on thin soil or bare wood; wrinkled leaves flattened along stem.

WAVY PLAGIOTHECIUM
Plagiothecium undulatum (Hedw.) B.S.G.

23b Plants not growing close to the ground.

24a Tangled and resembling excelsior (shredded wood); spreads over large areas in the rain forest; egg-shaped spore capsules are reddish brown with white peristome teeth.

LITTLE SHAGGY MOSS
Rhytidiadelphus loreus (Hedw.) Warnst.

24b Less tangled, and coarser; drier, more sunny areas.

BIG SHAGGY MOSS
Rhytidiadelphus triquetrus (Hedw.) Warnst.

25 Plants rope-like; wavy leaves thickly clustered along stems, not flattened.

ROBUST MOSS
Rhytidiopsis robusta (Hook.) Broth.

8b Plants growing in soil or rock.

26a Mosses growing in soil among grasses.

27a Wire-like mosses with tips of leaves bent back.

BENT-LEAF MOSS
Rhytidiadelphus squarrosus (Hedw.) Warnst.

27b Soft green mosses.

LAWN MOSS
Brachythecium albicans (Hedw. B.S.G.

26b Mosses growing on rock.

28a Mosses tree-like; scroll-shaped when dry, like the scroll of a violin.

BALSAM-TREE MOSS
Dendroalsia abietina (Hook.) Britt.

28b Mosses not tree-like.

29a Leaves flattened along the stem with a 'permanent wave', pale green.

30a Seta shorter than the capsule.

MENZIES'S NECKERA
Neckera menziesii Hook. ex Drumm.

30b Seta longer than the capsule.

DOUGLAS'S NECKERA
Neckera douglasii Hook.

29b Leaves not flattened.

31a Leaves large, soft and curly; tips have a drawn-out, stringy appearance; individual plants arise from stolons (underground stems).

CARDOT'S MOSS
Isothecium spiculiferum (Mitt.) Ren. & Card. var. *cardotii* (Kindb. ex Mac. & Kindb.) Crum, Steere & Anderson.

31b Rough raspy mosses, especially when dry; light green in color; prefer habitat that is wet at least part of the year.

ROUGH MOSS
Claopodium crispifolium (Hook.) Ren. & Card.

II. UPLAND MOSSES, above 3000 feet.

Moss plants upright; capsules, if present, on tips of stems (acrocarpous).

1a Habitats continuously wet: plants packed in thick beds along alpine streams and in alpine swamps.

2a Plants green.

3a Leaves short; some plants will have umbrella-like arms at tip of stem; fresh spore capsules apple-shaped.

4a Leaves twisted when dry.

AMERICAN CREEK MOSS
Philonotis americana Dism.

4b Leaves not twisted when dry.

FOUNTAIN CREEK MOSS
Philonotis fontana (Hedw.) Brid.

3b Leaves long; dense mat of red hair near the base.

RIBBED BOG MOSS
Aulacomnium palustre (Hedw.) Schwaegr.

2b Plants reddish; leaves oval; seldom fruiting.

BRYUM
Bryum species

1b Habitats not continuously wet; mainly dependent on current weather.

5a Plants on wood or in soil.

6a Bright-green leaves.

7a Large clumps of curly mosses on ground or on tree roots.

ALPINE CURLY HERON'S-BILL MOSS
Dicranum pallidisetum (Bailey ex Holz.) Irel.

7b Tiny mosses with leaves bent all in one direction; spore capsules bent over; in mats.

ALPINE BROOM MOSS
Kiaeria starkei (Web. & Mohr.) I. Hag.

7c Large oval-leaved moss; long yellow setae, and spore capsule as much as ¼ inch long.

SANDBERG'S BRYUM
Bryum sandbergii Holz.

6b Dull-green stubby mosses; dry plant, egg-shaped.

8a Leaves, hair-tipped; square-angled capsule.

AWNED HAIRCAP
Polytrichum piliferum Hedw.

8b Leaves not hair-tipped; capsule cylindrical.

URN BEARDED MOSS
Pogonatum urnigerum (Hedw.) P. Beauv.

5b Plants on rocks.

9a Mosses in button-like clusters.

10a Leaves short.

ALPINE GRIMMIA
Grimmia alpestris (Web. & Mohr) Schleich.

10b Leaves long.

ROCK STRAIGHT-HAIRED MOSS
Orthotrichum rupestre Schleich. ex Schwaegr.

9b Mosses not in button-like clusters.

11a Very small mosses, in thin layers.

12a Plants red.

RED ROCK MOSS
Andreaea nivalis Hook.

12b Plants black.

BLACK ROCK MOSS
Andreaea rupestris Hedw.

11b Larger mosses, in thicker layers.

13a Leaves curled.

WEIS DICRANUM
Dicranoweisia crispula (Hedw.) Lindb. ex Milde

13b Leaves not curled.

GRAY FRAYED-CAP MOSS
Rhacomitrium canescens (Hedw.) Brid.

B Moss plants creeping; spore capsules if any, along stems (pleurocarpous).

1a Habitats continuously wet: thick beds of light-green moss, either in fast water or splashed with water for the greater part of the year.

2a Leaves large; plants thick and cushiony; tips of branches sharp and shiny.

GOLDEN SHORT-CAPSULED MOSS
Brachythecium asperrimum (Mitt.) Sull.

2b Leaves small, plants like fine ferns.

3a Leaves not curled.

SLENDER BEAKED MOSS
Eurhynchium praelongum (Hedw.) B.S.G. var. *stokesii* (Turn.) Hook.

3b Leaves curled.

4a Paraphyllia abundant (small green growths of various shapes) between leaves.

HOOKED MOSS
Cratoneuron filicinum (Hedw.) Spruce

4b Paraphyllia absent.

SICKLE MOSS
Drepanocladus uncinatus (Hedw.) Warnst.

1b Habitats not continuously wet; mainly dependent on current weather.

5a Mosses in open timber and on alpine meadows.

6a Mosses in litter of forest floor.

7a Large, coarse plant with blunt tip turned back over itself; tips yellowish green; bases brown.

GIANT MOSS
Homalothecium megaptilum (Sull.) Robins.

7b Coarse, rope-like moss in dense patches; wavy leaves thickly clustered along stems, not flattened.

ROBUST MOSS
Rhytidiopsis robusta (Hook.) Broth.

7c Light-yellowish moss, straw-like, in loosely packed beds; stems reddish.

SCHREBER'S MOSS
Pleurozium schreberi (Brid.) Mitt.

6b Mosses not in litter of forest floor.

8a Mosses on the bases of trees.

9a Bright-green mosses; flattened along the stem.

ELEGANT MOSS
Isopterygium borrerianum (C. Mull.) Lindb.

9b Dull-yellowish-green mosses; not flattened.

LESQUEREUX'S MOSSES
Lescuraea species.

8b Mosses on rock or in soil.

10a Soft, light-green mosses.

LAWN MOSS
Brachythecium albicans (Hedw.) B.S.G.

10b Dull-yellowish-green mosses.

LESQUEREUX 'S MOSSES
Lescuraea species.

SPECIES DESCRIPTIONS AND DRAWINGS

Anacolia menziesii Bartramiaceae

RED-HAIRED MOSS

One will find *Anacolia* growing on bare knobs of rock that stand out along valley sides. It belongs to the same family as *Philonotis* and *Bartramia.* If you are fortunate, you may find it in fruit, with the apple-shaped spore capsules typical of the family. The red tomentum identifies this species.

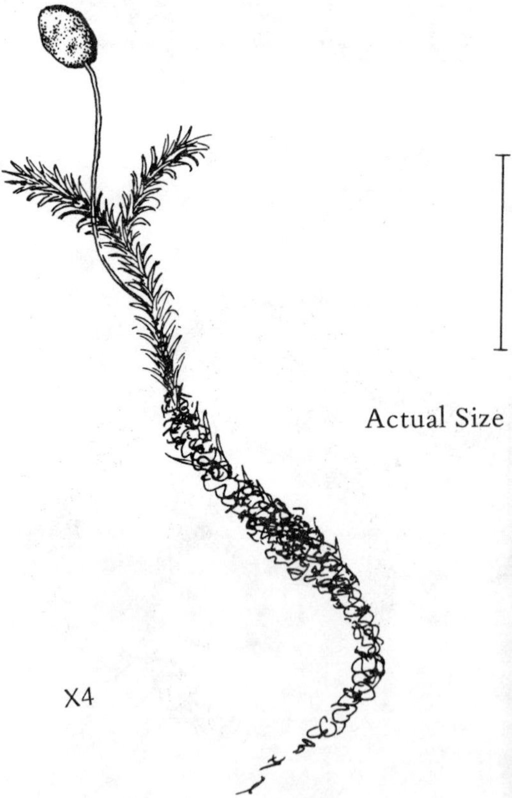

Actual Size

X4

Andreaea nivalis **Andreaeaceae**

RED ROCK MOSS

Largest of all North American Rock Mosses, *Andreaea nivalis* is always reddish brown or darker (but never black), soft to the touch, not harsh or brittle.

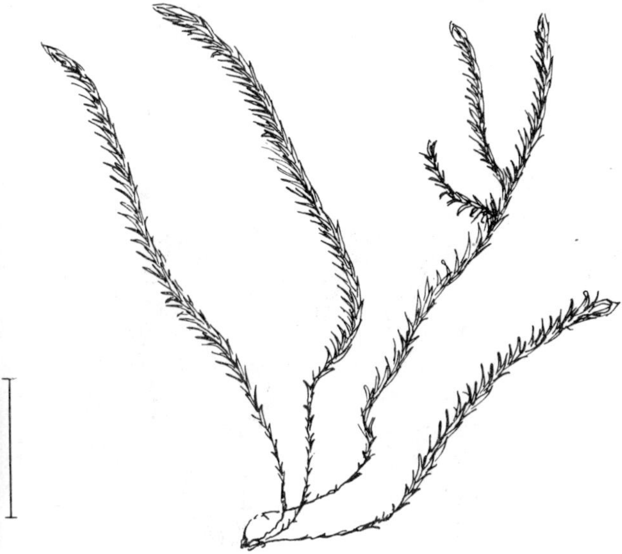

X4

Andreaea rupestris **Andreaeaceae**

BLACK ROCK MOSS

For one who loves the primitive environment of the high mountain peaks, these tiny mosses are fit companions. Lichen-like, they cling to the bare rocks of those crags that jut above timberline. Pioneers in soil making, the Andreaeas are also pioneers in the evolutionary development of the mosses. The spore capsule opens by four slits to free the spores.

X7

I

Antitrichia curtipendula **Chryphaeaceae**

HANGING MOSS

Common in the wet lowlands, this moss is most likely to be observed in the rain forest. Typically found hanging from limbs, but often picked up on the ground where it has fallen during storms; a hand lens will reveal that the midrib or costa has a short costa on either side.

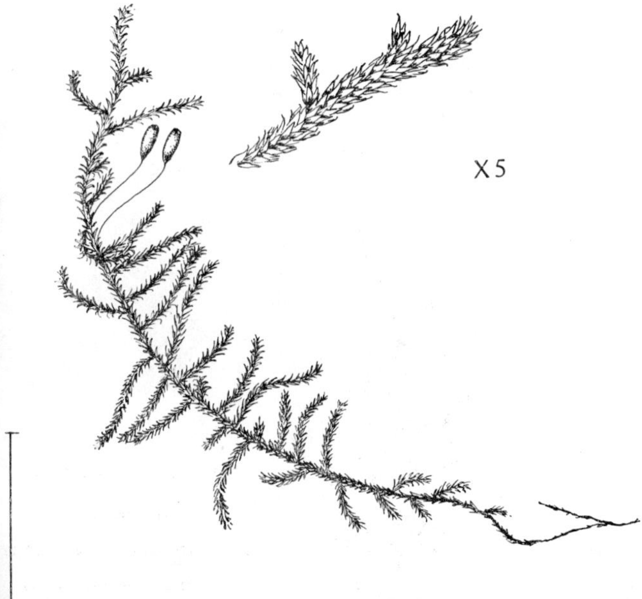

Atrichum undulatum **Polytrichaceae**

QUEEN CATHERINE'S MOSS

A 'Mexican Hairless' among the haircaps. The name *Atrichum* means without hair. As with *Tetraphis,* this moss was named after a royal patron, Empress Catherine II of Russia. With the adoption of standard rules of nomenclature, the euphonious *Catherinia* had to give way to *Atrichum.* In order to preserve the name, we will call it Queen Catherine's Moss. The wavy leaves of this species make it one of the prettier mosses.

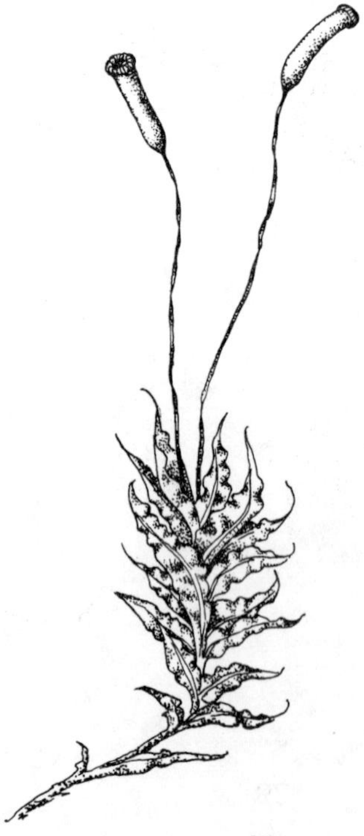

X3

Aulacomnium androgynum **Aulacomniaceae**

LOVER'S MOSS

An interesting little species because of its stalked gemmae or fruiting bodies. These are produced in the spring and persist until early summer. Even when the yellow spore capsules with their white peristomes appear, the headless stalks may still be found.

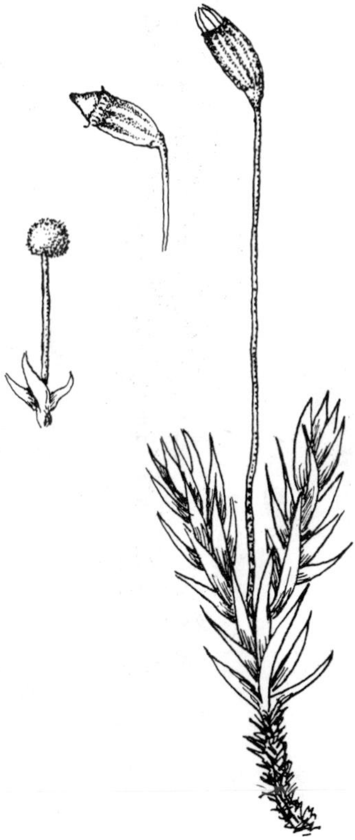

X2.5

Aulacomnium palustre **Aulacomniaceae**

RIBBED BOG MOSS

As its specific name signifies, this moss is a denizen of the bogs and marshes. Boasting red 'hair', as does *Anacolia,* it is easily identified. The generic name means 'furrowed' or 'ribbed', referring to the dried capsules.

X2

Bartramia pomiformis var. *elongata* **Bartramiaceae**

APPLE MOSS

The true apple moss is more common in the eastern states. With its apple-round spore capsules and soft, curled leaves, our moss is a lovely little plant. It is common on the Mount Storm King trail, in Olympic National Park.

X4

Brachythecium albicans **Hypnaceae**

LAWN MOSS

Brachythecium means 'short capsule'. Few mosses have been given common names, and it would be better to disregard these and concentrate on learning their botanical names. Although this moss may be found on the ocean beaches, the high mountain meadows, and the prairies of eastern Washington, it is best known as the moss that takes over our lawns when we fail to fertilize them.

On the bare rocks of the highest peaks, crustose lichens take over the soil-making process. Soon, mosses with their ability to live on sterile soil move in, and virgin rock is slowly made into soil to support flowering plants.

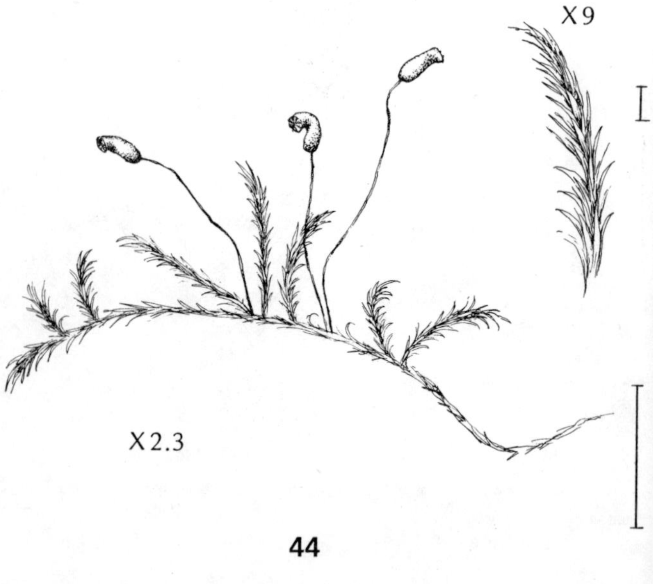

Brachythecium asperrimum **Hypnaceae**

GOLDEN SHORT-CAPSULED MOSS

On most waterfalls, and places where trails cross running water, the rocks will be cushioned with this moss.

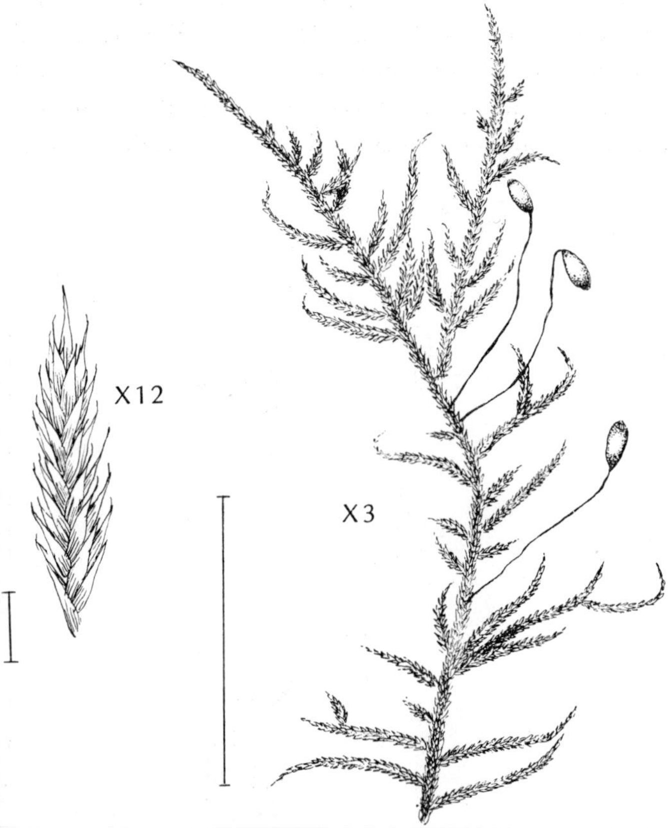

Bryum species **Bryaceae**

Bryum sandbergii

SANDBERG'S BRYUM

The Bryums are easy to recognize with their pendulous, pear-shaped capsules, and their leaves in a basal rosette. The species may be difficult to determine. Few people get to see Sandberg's Bryum, as it is at home only at mile-high elevations. Preferring soil in alpine groves, this moss passes for a Mnium until you can see its large, yellow spore capsules, almost a quarter of an inch long, and its long, yellow setae.

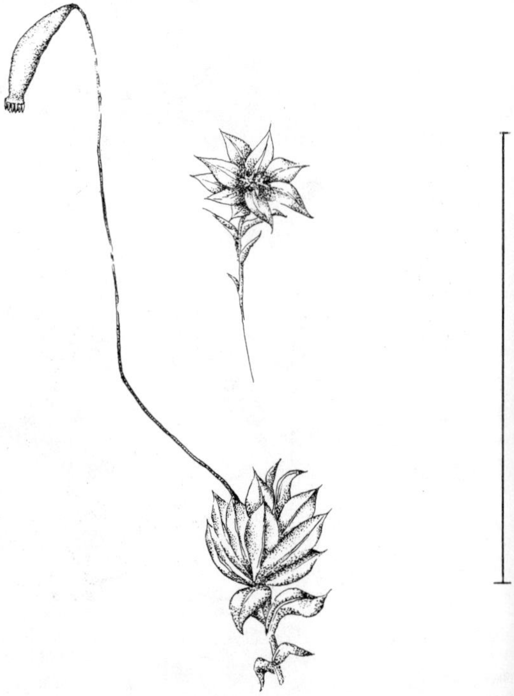

X1.5

Ceratodon purpureus **Ditrichaceae**

PURPLE HORNTOOTH

Purple Horntooth is of worldwide distribution. It can be recognized in the spring and early summer by the erect, purplish capsules, which are deeply grooved. Observation will show how the peristome teeth close over the mouth of the capsule in wet weather, protecting the spores. As the plant dries, the teeth open and spores are caught by the wind to be distributed all over the earth.

X2.5

Claopodium crispifolium **Leskeaceae**

ROUGH MOSS

A peculiar, rough look sets this large moss apart from its neighbors. Examination under a compound microscope confirms this distinction. Each leaf cell is covered with rough lumps or papillae. The leaf has a long, twisted-hair point and is contorted in general appearance.

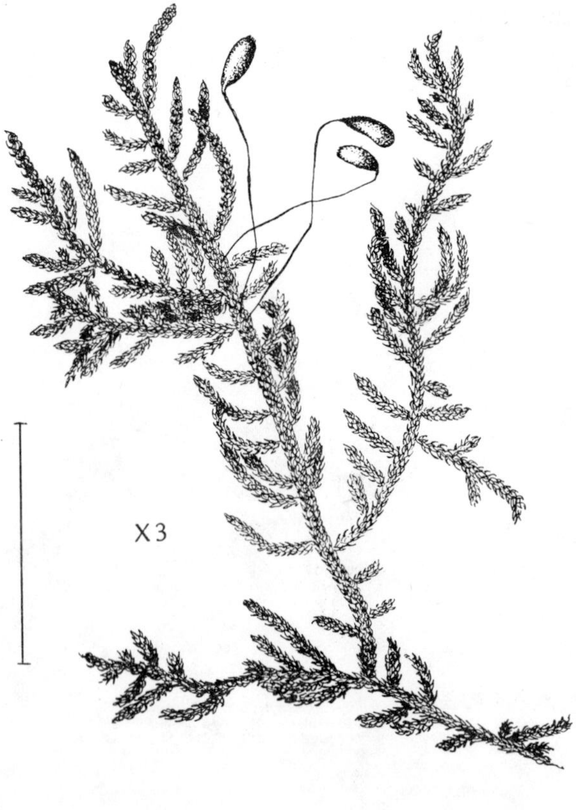

Climacium dendroides Hypnaceae

TREE MOSS

Another tree moss found in swamps. *Climacium* refers to a staircase, and *dendroides,* of course, is Greek for tree-like.

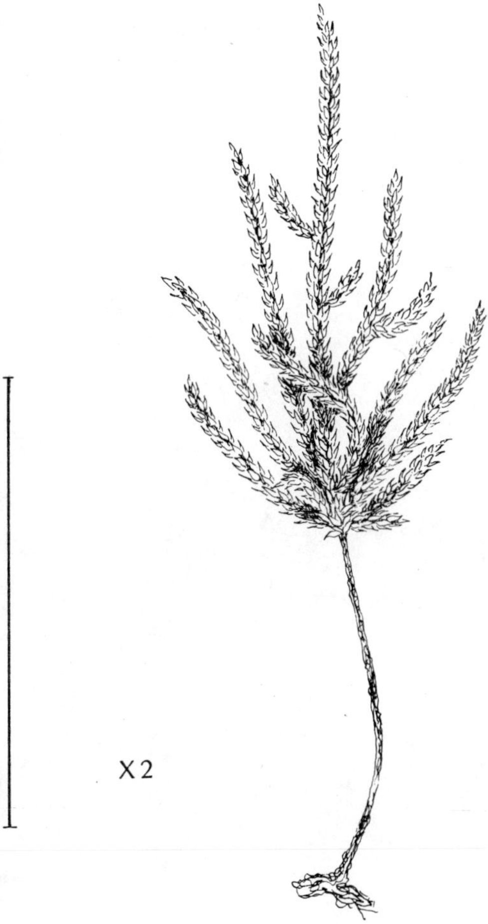

X2

Cratoneuron commutatum var. *falcatum* **Hypnaceae**
Cratoneuron filicinum
HOOKED MOSS

The Cratoneurons are not common. A hand lens will be necessary in order to see the paraphyllia between the leaves.

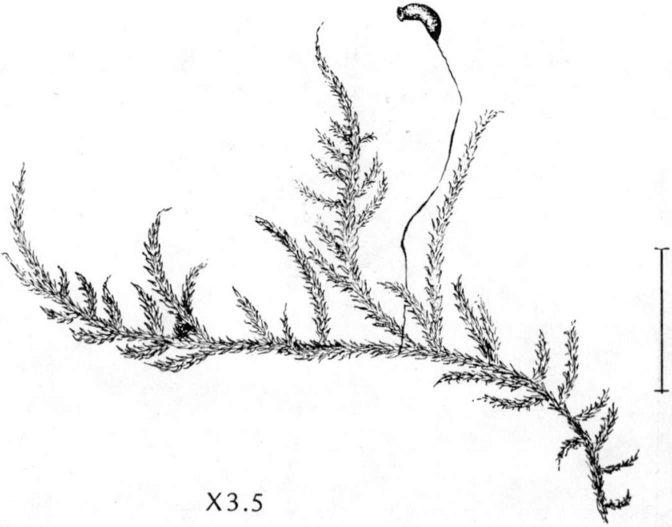

X3.5

Dendroalsia abietina Cryphaeaceae
BALSAM-TREE MOSS

Rather rare, but very distinctive when you see it. Another tree moss; but when it is dry the top curls in on itself like the scroll of a violin.

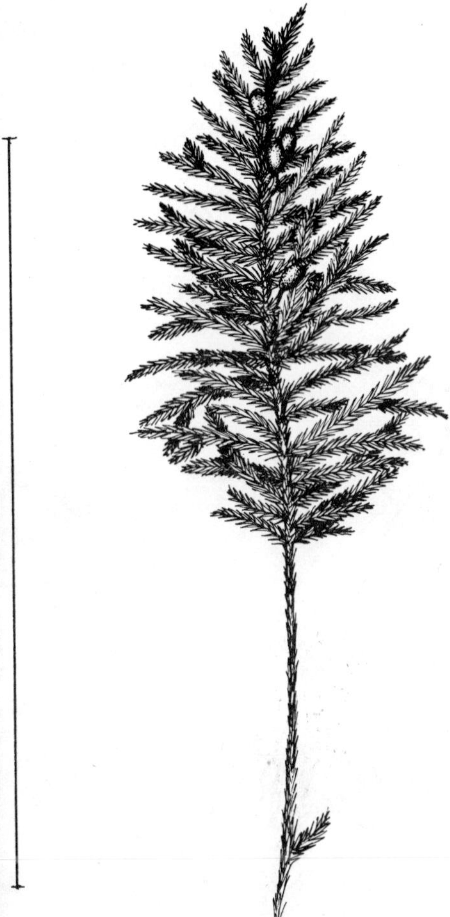

X1.25

Dicranoweisia crispula **Dicranaceae**

Dicranoweisia cirrhata

WEIS DICRANUM

With the exception of this one genus, all the members of the Dicranaceae have split peristome teeth, as do the members of the Ditricaceae. In the Ditricaceae, the teeth are split nearly to the base, and in the Dicranaceae they are split slightly more than half way down. In some species the peristomes are single, and in others there is an inner peristome, more or less transparent, with cilia in between. The species *cirrhata* may be distinguished from *crispula* by the general fact that *cirrhata* grows mainly on wood, while *crispula* confines its habitat primarily to rock.

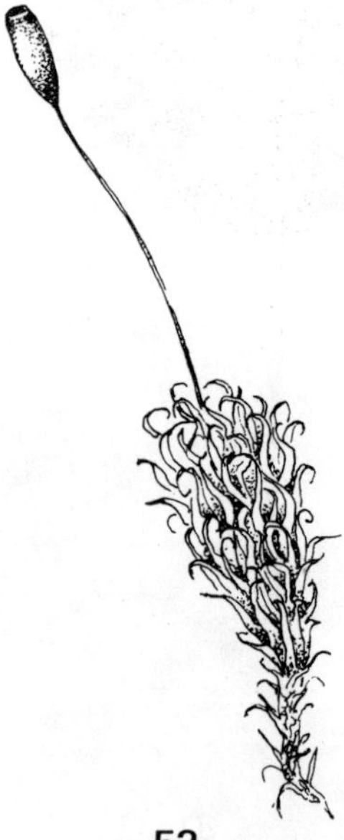

X4

Dicranum fuscescens **Dicranaceae**

CURLY HERON'S-BILL MOSS

Very common on tree trunks; a very similar species, *D. pallidisetum*, **Alpine Curly Heron's-Bill Moss,** may be found growing on the ground near timberline.

X3

Dicranum scoparium **Dicranaceae**

HERON'S-BILL or BROOM MOSS

Same habitat as the preceding: common in river bottoms.

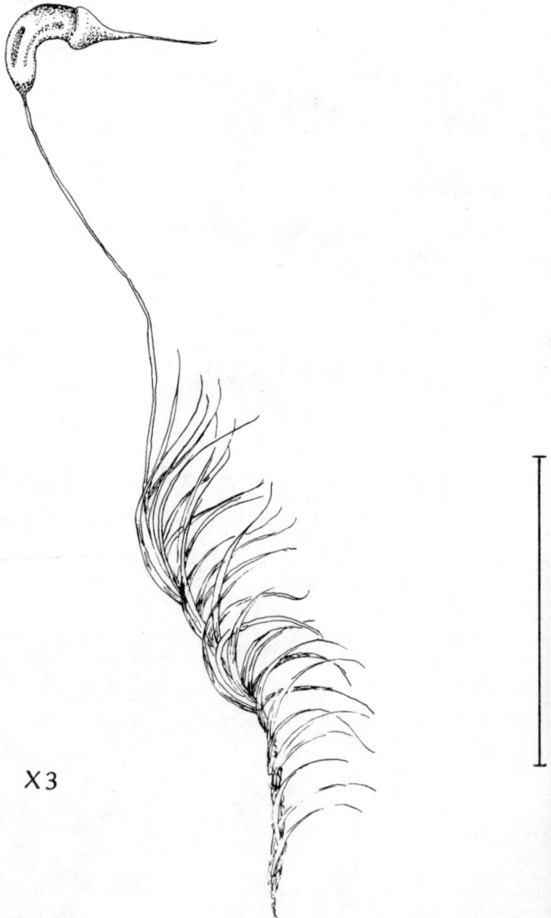

X3

Dicranum tauricum **Dicranaceae**

BROKEN-LEAF MOSS

An easily identified moss; most commonly seen on fallen logs, in flat, yellowish-green clumps. A hand lens will reveal that almost all of the leaves are broken off. Differs from its relatives in its erect spore capsules.

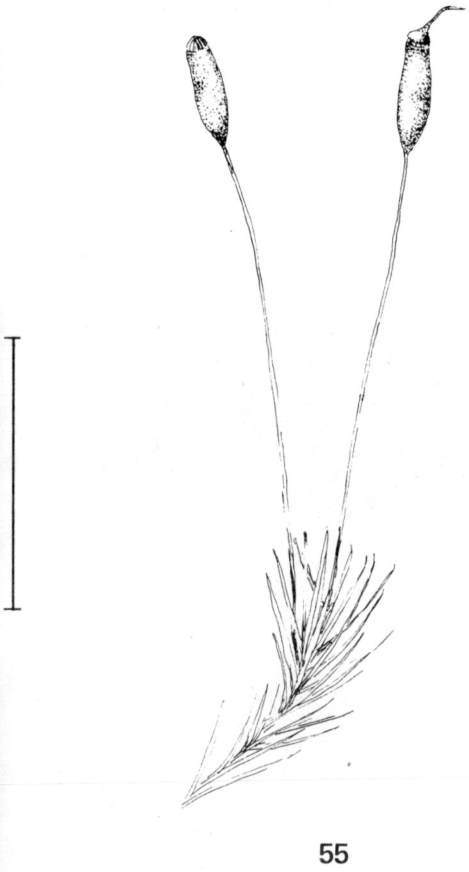

X3

Drepanocladus aduncus **Hypnaceae**
Drepanocladus uncinatus
SICKLE MOSS

The Sickle Mosses are plants of the swamps and ponds. The genus has a great many species, difficult to discriminate. A similar genus, *Cratoneuron,* may be recognized by paraphyllia on the stem between the leaves. These structures may be hair-like, leaf-like, or knife-like. They may be seen with a strong hand lens. The species *uncinatus* is more likely to be found in the higher, more alpine terrains.

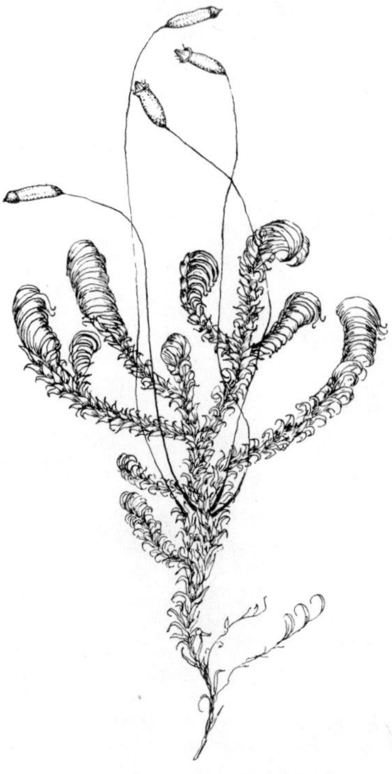

X3

Encalypta rhaptocarpa **Encalyptaceae**

EXTINGUISHER MOSS

The author has found these mosses under overhanging banks. They are not common. With the long skirt-like calyptra they can be mistaken for no other moss. The common name comes from the similarity of the calyptra to a candle extinguisher.

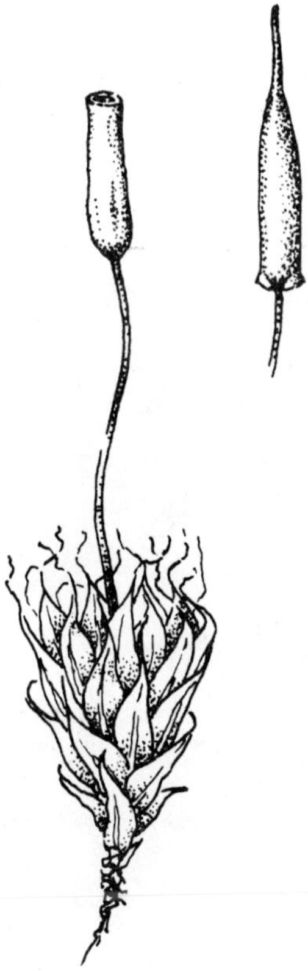

X2.5

Eurhynchium oreganum **Hypnaceae**

OREGON BEAKED MOSS

One of the outstanding mosses of the Pacific Coast. Its large, feather-like fronds cover the whole forest floor in places. Varying in color from green to a deep gold, it is truly spectacular, and helps to make the region a bryologist's paradise. The common name comes from the fact that the cap (operculum) over the peristome teeth has a long beak. Covering the entire top of the capsule when it is immature is the calyptra, which is the first structure to be shed; it is the remains of the archegonium or case that enclosed the egg.

X3/4

Eurhynchium praelongum var. *stokesii* **Hypnaceae**

SLENDER BEAKED MOSS

Similar to the Oregon Beaked Moss, but finer, and more likely to be found on wet rocks along streams.

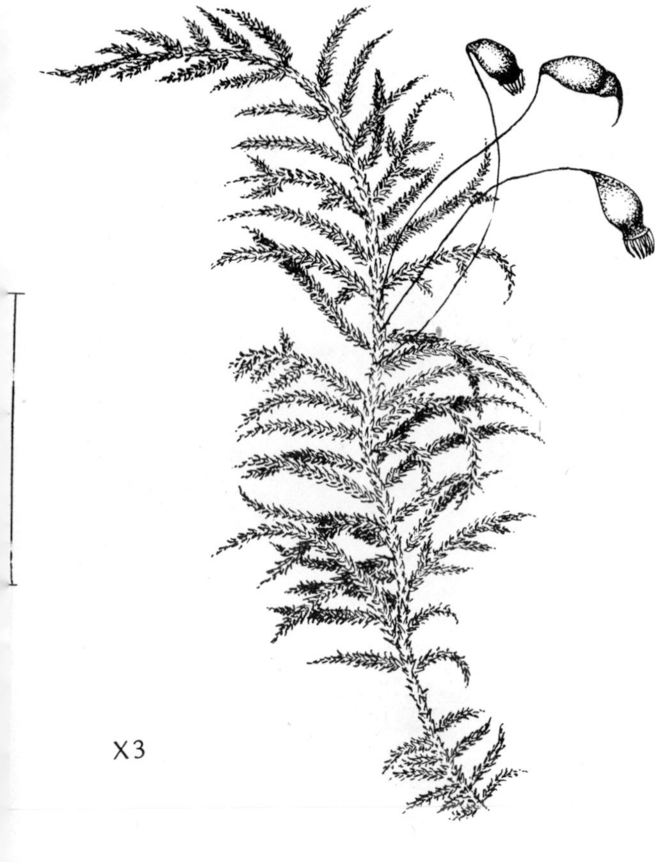

X3

Fontinalis antipyretica var. *gigantea* **Fontinalaceae**

GIANT FOUNTAIN MOSS

Not too common. The author found it in swiftly flowing streams near timberline. Strands may be several feet long, with leaves up to a half-inch in length.

X1.5

Fontinalis neomexicana Fontinalaceae

NEW MEXICAN FOUNTAIN MOSS

More common. This moss, although but a miniature of the preceding, is still a large moss. It may be found growing in creeks that flow through river bottoms.

Funaria hygrometrica **Funariaceae**

TWISTED-CORD MOSS

Another moss of universal distribution. Commonly found on the sites of old campfires.

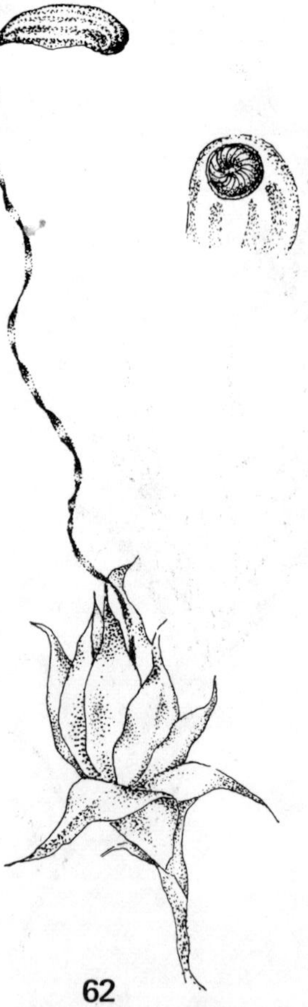

X4

Grimmia alpestris **Grimmiaceae**

ALPINE GRIMMIA

This Grimmia is found only on the higher peaks. Shapely buttons of dark green, covered with mouse-gray leaf tips, bear an abundance of small, light-brown spore capsules.

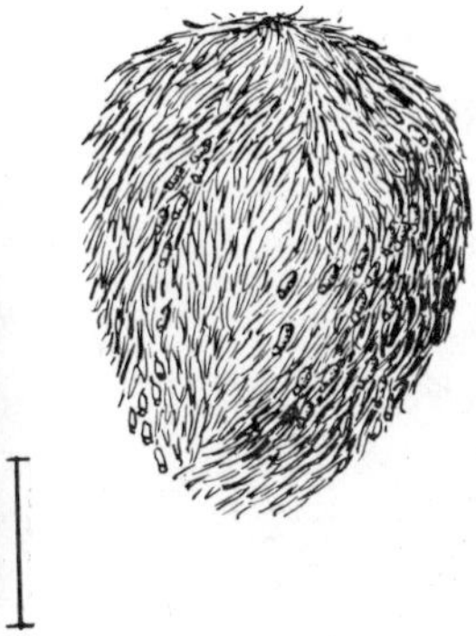

X3

X6

Grimmia alpicola **Grimmiaceae**

COMMON GRIMMIA

The Grimmias are associated with rocks. Growing under water at flood stage, as does *G. alpicola*, or forming buttons on the highest mountain peaks, they follow the lichens in nature's soil-forming succession.

X4

Grimmia maritima **Grimmiaceae**

OCEAN GRIMMIA

A typical button Grimmia, found only on cliffs facing the ocean. For a Grimmia, the leaves are large. Their twisted shape, when dry, readily identifies them. Look for it only on rocks in reach of the salt spray.

X10

Homalothecium fulgescens **Hypnaceae**

YELLOW MOSS

A moss that you will learn to recognize by general appearance. A hand lens will reveal long, pointed leaves with lengthwise folds.

X3

Homalothecium megaptilum **Hypnaceae**

GIANT MOSS

Another moss of the open Canadian Zone timber. Fronds may be as much as four inches long. Only the upper half shows the yellowish-green sheen that makes it so easy to see on the forest floor.

X2.5

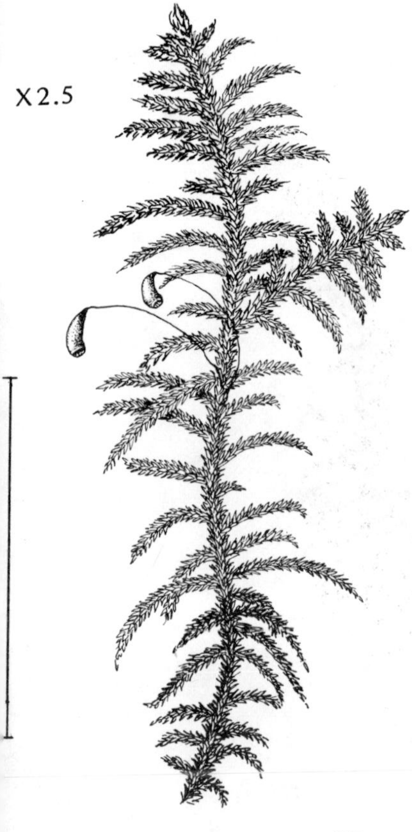

X10

Homalothecium nuttallii **Hypnaceae**

CREEPING BEAUTY

This may well be called Creeping Beauty, as the narrow, graceful fronds wind over the smooth white bark of deciduous trees. It isn't always found on tree trunks. The author found his first specimen on a concrete post in the Queets rain forest.

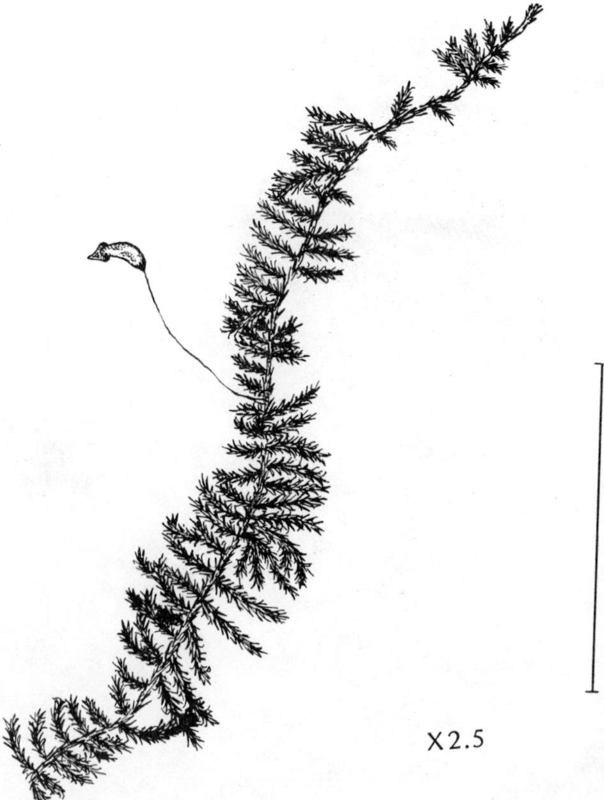

X2.5

Hylocomium splendens **Hypnaceae**

MOUNTAIN FERN MOSS

Another of the spectaculars among the Pacific Coast mosses. Whole banks and hillsides will be covered with the multiple fronds of this moss. The age (in years) of this interesting moss may be easily told by counting the number of times a plant has grown out of the one next below it.

X2/3

Hypnum circinale **Hypnaceae**

COILED-LEAF MOSS

This pretty little moss is quite common on the trunks of Western Hemlock trees. The delicate branches all seem to flow down hill, giving an easily recognized appearance.

X4

Hypnum subimponens **Hypnaceae**

CURLY HYPNUM

This moss with its tightly curled fronds varies in color from green to gold. Mosses tend to have a golden color in the late summer, especially in the rain forest.

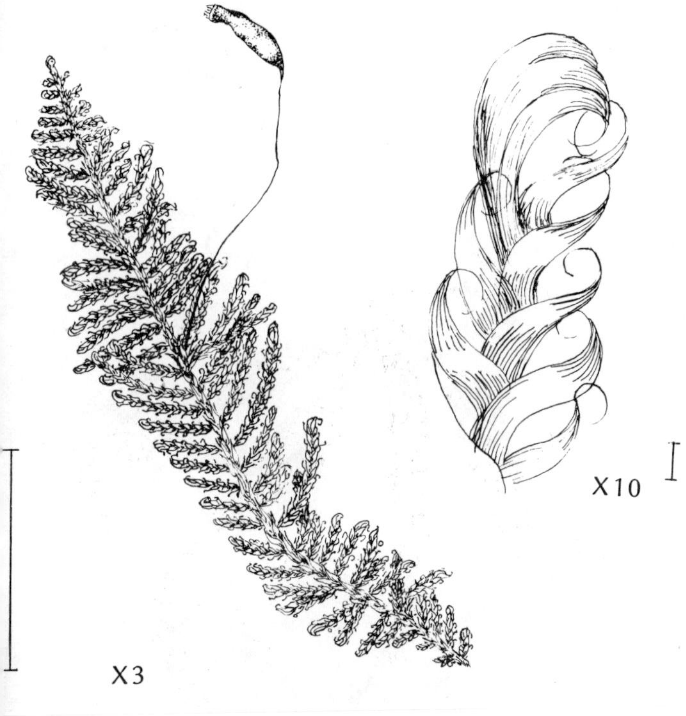

Isopterygium borrerianum **Hypnaceae**

ELEGANT MOSS

A scintillating little plant that loves to abide alone on its substratum, which generally consists of a tree root. As with most of this family, the leaves are complanate (lie in one plane along the stem).

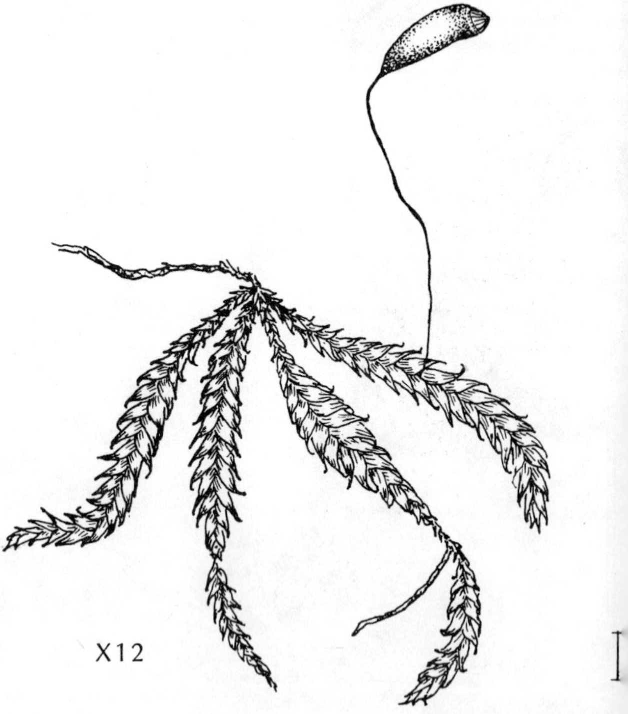

X12

Isothecium spiculiferum var. *cardotii* **Hypnaceae**

CARDOT'S MOSS

The large mosses that you have seen covering many tree trunks and large rocks are exactly the same species; every gradation in leaf size exists from the gigantic Cardot's Isothecium to the delicate flagellate form hanging from the limbs.

Isothecium spiculiferum **Hypnaceae**

LACE MOSS

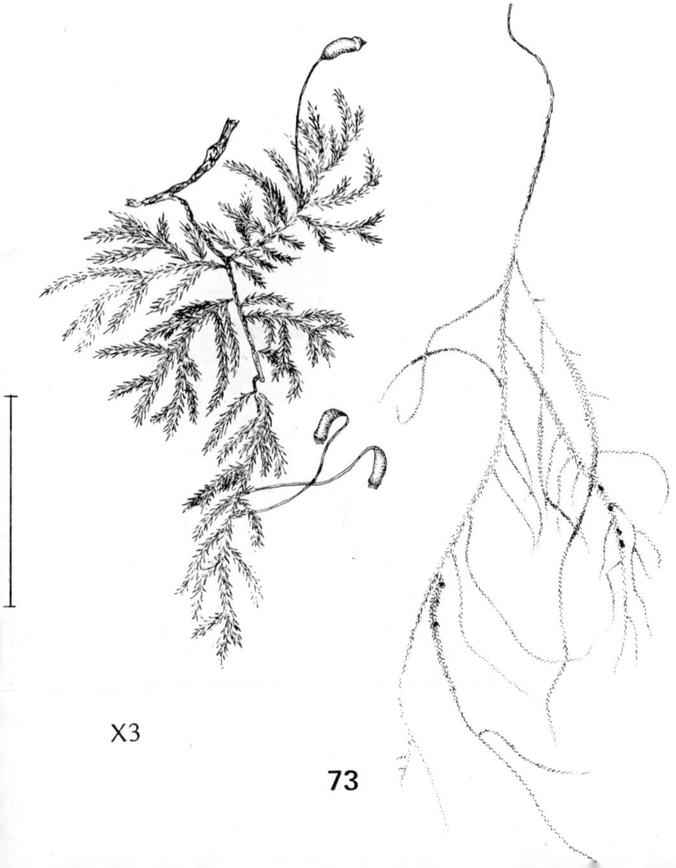

X3

Kiaeria starkei **Dicranaceae**

ALPINE BROOM MOSS

One needs a compound microscope to positively identify the Kiaerias. However, the chances are fairly strong that if you have a tiny Broom Moss above 4000 feet it is of this species.

X5.5

Lescuraea species **Leskeaceae**

LESQUEREUX'S MOSSES

These mosses derive their generic name from an early American bryologist, Leo Lesquereux. The species are very hard to discriminate.

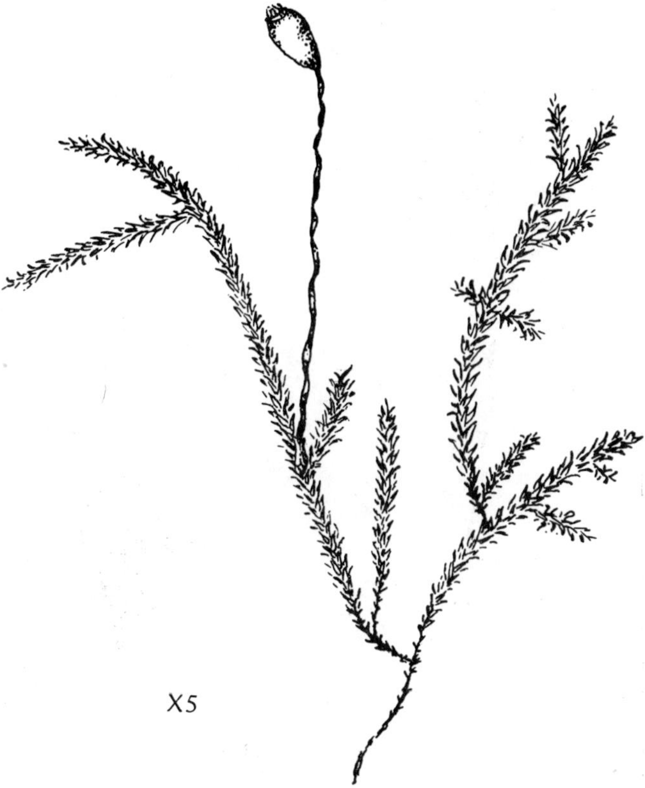

X5

Leucolepis menziesii **Mniaceae**

MENZIES'S TREE MOSS

Another of our prettiest mosses. Each individual plant looks like a tiny tree. This is one of several mosses that have this tree-like habit of growth.

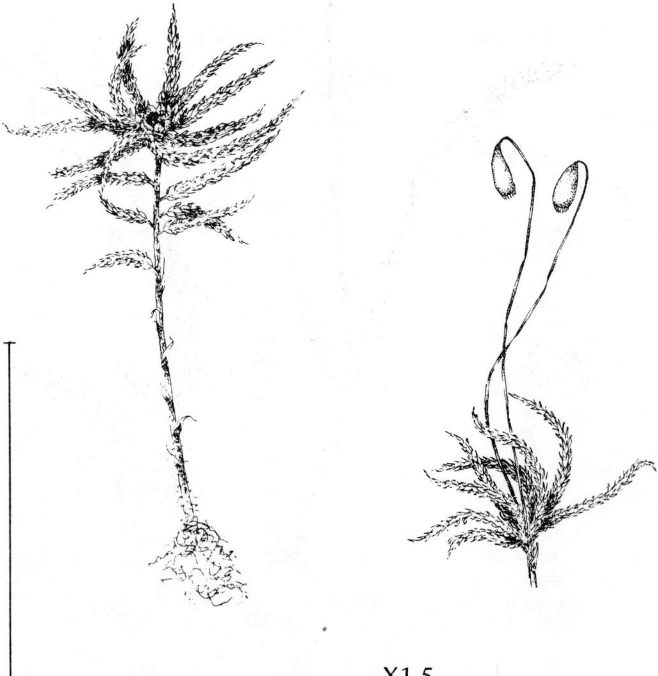

X1.5

Mnium insigne

BADGE MNIUM

In wet weather the sterile plants of this species look like tiny ferns. In dry weather they shrink up and are not noticed. One reason that a landscape 'freshens up' so quickly after a rain is the rapidity with which a dry moss plant can absorb water and resume its activity. Pour a few drops of water from your canteen on some dry moss, on a hot August hike, and notice how quickly it will resume its April appearance. The fruiting plants of *insigne* resemble those of *punctatum*, only they are much larger.

Mnium punctatum

EARLY MNIUM

Mnium is a Greek name for moss. Because of their large, thin leaves, the Mniums are favorite laboratory mosses. Even a twenty-power hand lens will reveal the cellular structure of the larger leaves.

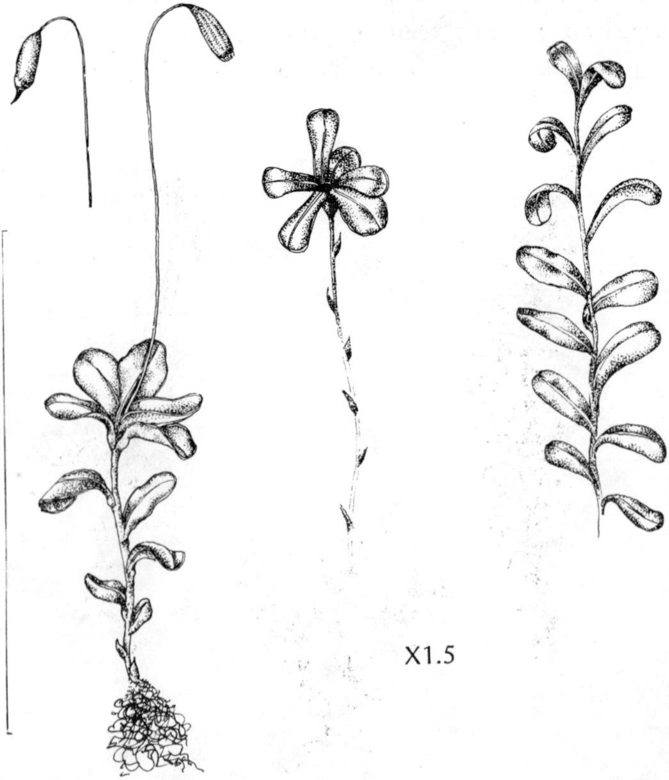

X1.5

Mnium spinulosum **Mniaceae**

RED-MOUTHED MNIUM

A typical Mnium, quite distinctive with its red-mouthed peristome.

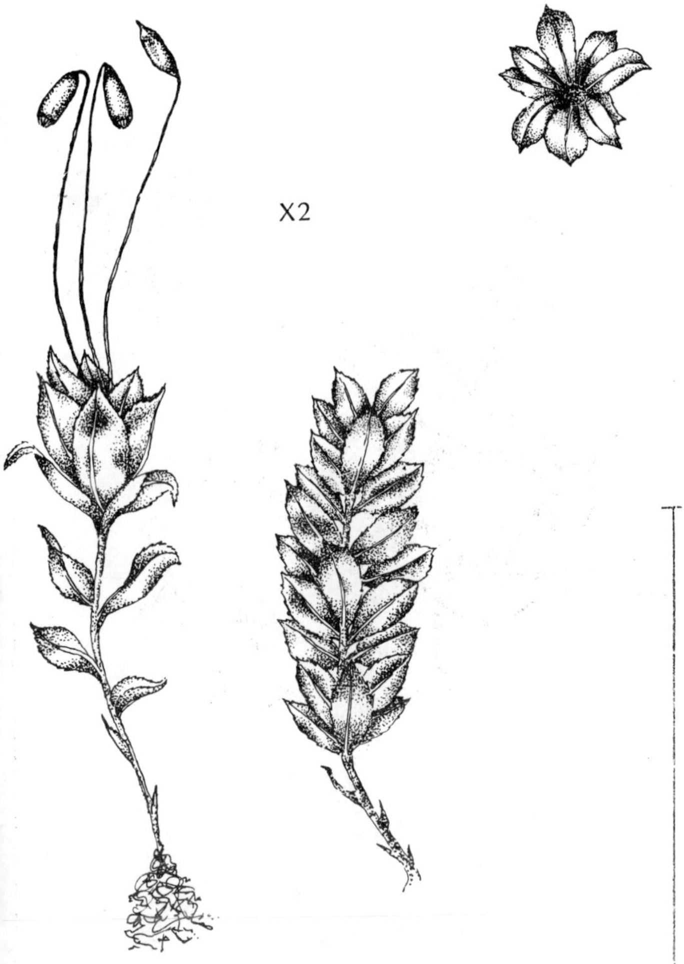

Neckera douglasii **Neckeraceae**

DOUGLAS'S NECKERA

The Neckeras are at their best in the watersheds draining into the ocean. However, they are widespread and form a conspicuous part of the moss flora.

X8

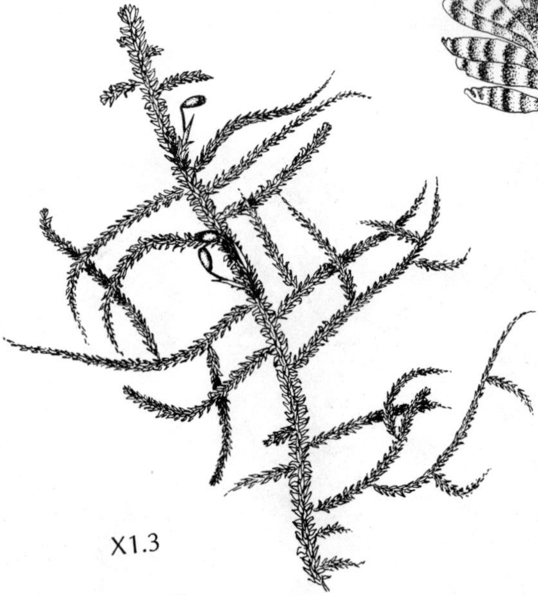

X1.3

Neckera menziesii **Neckeraceae**

MENZIES'S NECKERA

An interesting account of David Douglas and Archibald Menzies may be found in Jones's *Botanical Survey of the Olympic Peninsula.*

X1.7

X10

Orthotrichum lyellii **Orthotrichaceae**
Orthotrichum rupestre
STRAIGHT-HAIRED MOSS

Another difficult genus: Besides the leaves, one needs old spore capsules, as well as fresh ones, for positive identification. The species *rupestre* is a little moss occurring in tiny buttons on rocks in the high country. The species *lyellii* is a much larger plant, common on broad-leaved trees of the rain forest.

X2

Philonotis americana **Bartramiaceae**

AMERICAN CREEK MOSS

Philonotis fontana

FOUNTAIN CREEK MOSS

The two main species are easily determined. However, the varieties are almost impossible to discriminate. Other species are relatively easy to key out. *Philonotis* means water-loving, and again, as in *Anacolia*, the apple-shaped spore capsules betray their family. Very common in alpine meadows.

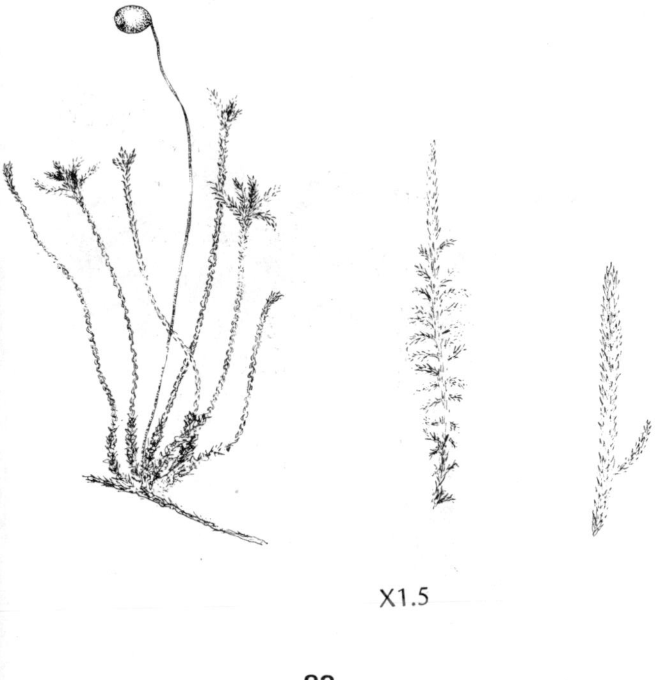

X1.5

Plagiothecium undulatum **Hypnaceae**

WAVY PLAGIOTHECIUM

Common on the trunks of fallen trees and the bases of trees. On some mountain trails, the author has seen it form sheets, with the stems all lying or 'running' down hill.

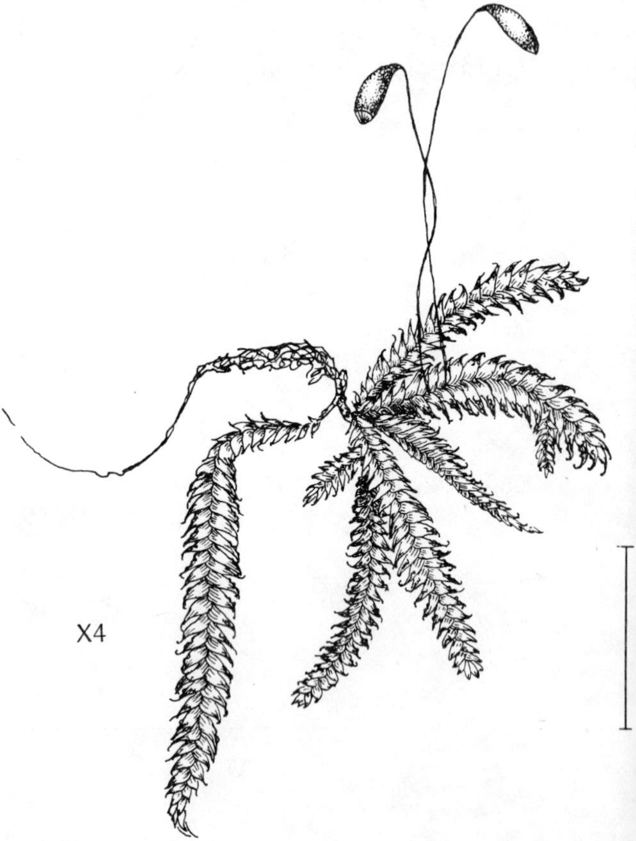

X4

Pleurozium schreberi **Hypnaceae**

SCHREBER'S MOSS

This moss is quite common in the open woods of the Canadian Zone of the drier mountain sides. As one hikes up the trail, the banks are quite likely to be covered with thick beds of this distinctive species.

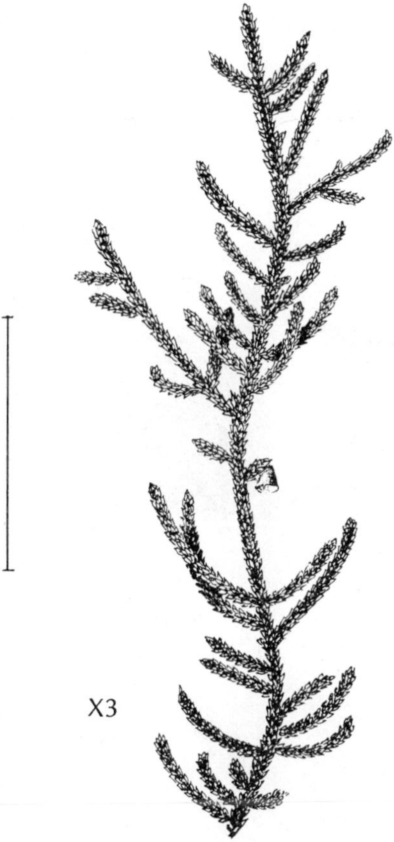

X3

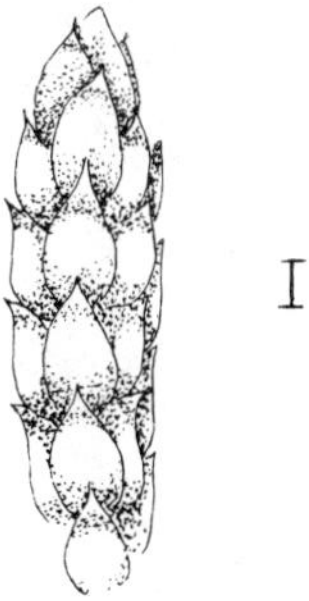

X12

Pogonatum alpinum **Polytrichaceae**

BEARDED MOSS

The members of this family are also called Pigeon Wheat. The Pogonatums are most likely to be found in moist, rich soil. The hair-cap comes from the long hairs that cover the calyptra in several genera. *Pogonatum* refers to a goat-like beard.

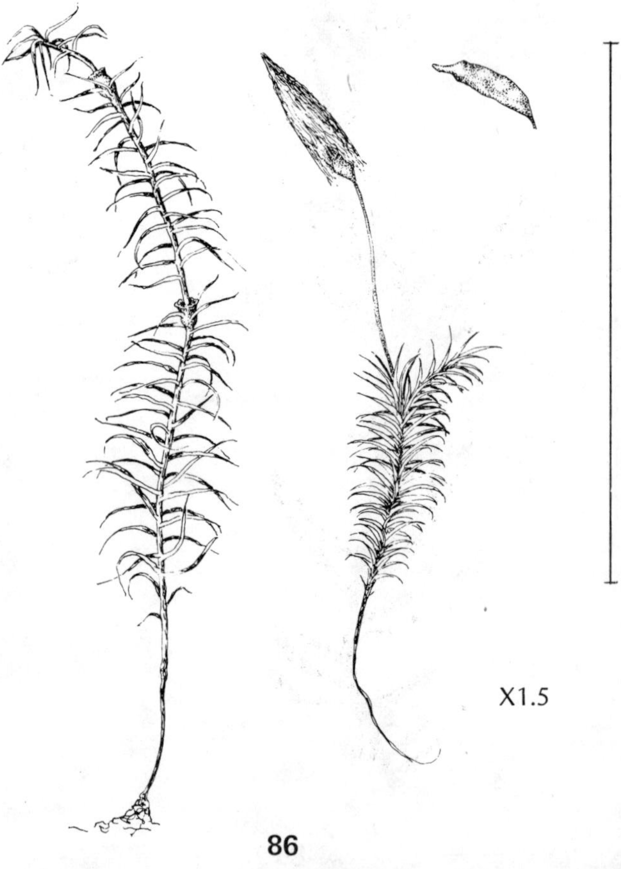

X1.5

Pogonatum urnigerum **Polytrichaceae**

URN BEARDED MOSS

Although also found near sea level, this stubby little moss is more common in alpine meadows. Apt to be mistaken for the Awned Haircap; close examination will reveal that the leaves do not have hair-like tips.

Pohlia nutans **Bryaceae**

NODDING POHLIA

The Pohlias are similar to the Bryums, and are almost as difficult to identify.

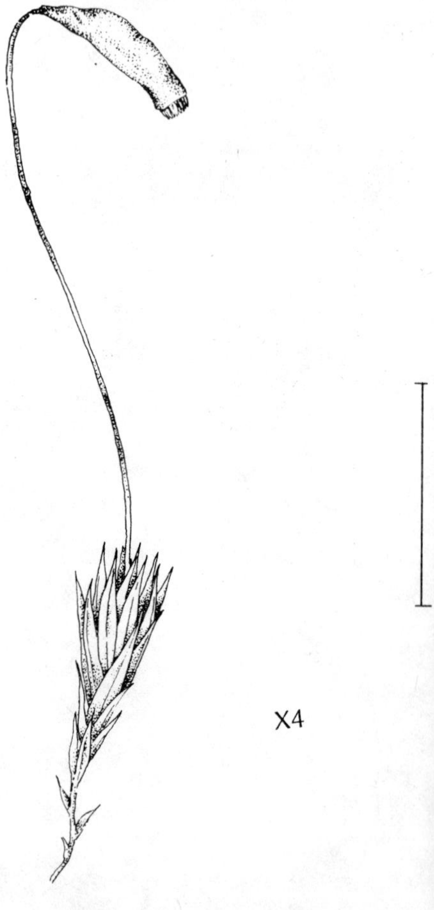

X4

Polytrichum juniperinum **Polytrichaceae**

JUNIPER HAIRCAP

One of our largest land mosses. Exceeded only by the Common Haircap, *Polytrichum commune,* which the author has found only near Point Grenville, Washington.

X1.5

Polytrichum piliferum
AWNED HAIRCAP

Small, stubby plant, like *Pogonatum urnigerum,* but having a hair tip on leaf. The spore capsules of this moss tend to be square-angled, and the species grows generally in the high mountain meadows. Many of the generic names of this family center around the hairs on the calyptra. *Polytrichum* means many hairs, *Atrichum,* no hairs, *Oligotrichum,* few hairs.

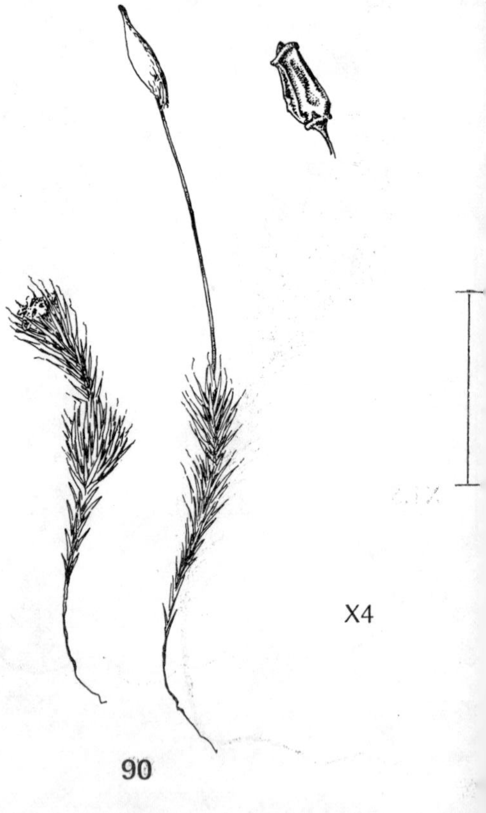

Rhacomitrium canescens var. *ericoides*
GRAY FRAYED-CAP MOSS

The Rhacomitriums are mosses that endure the most drying because of their living on rocks. Let a little rain fall on them and they quickly come to life. The gray color of this species is contributed by the thousands of tiny, transparent leaf points.

X3.5

Rhacomitrium varium **Grimmiaceae**

GREEN FRAYED-CAP MOSS

There are more Rhacomitriums, but these two are easily recognized by the beginner. If the species that are not listed bother you, then you are sufficiently interested to get a standard work on mosses and commence your study seriously.

X4

Rhytidiadelphus loreus **Hypnaceae**

LITTLE SHAGGY MOSS

In sheer bulk, this moss is the commonest moss in the forested areas. The rain forest is noted for its mosses; however, it is not that there are so many different kinds of moss but that there are tremendous quantities of several common kinds, such as Little Shaggy.

X1

X5

Rhytidiadelphus squarrosus **Hypnaceae**

BENT-LEAF MOSS

The generic name of these plants refers to a 'wrinkled brotherhood', probably because the leaves are more wrinkled than in most genera.

X7

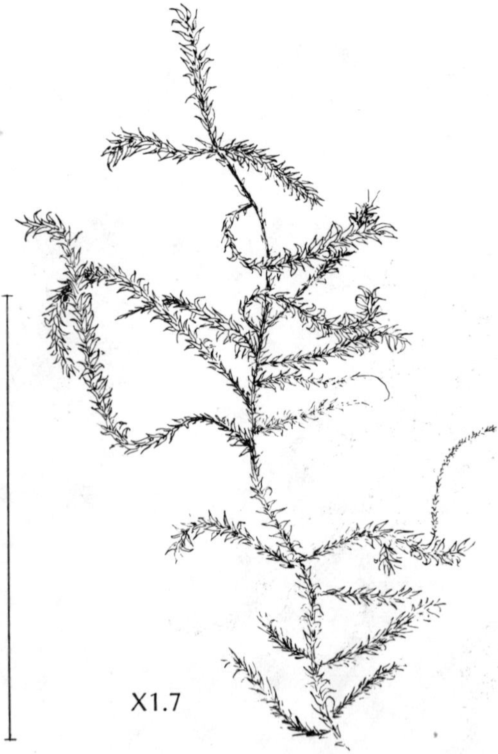

X1.7

Rhytidiadelphus triquetrus **Hypnaceae**

BIG SHAGGY MOSS

One of the big mosses, not common, but widespread. Easily confused by the beginner with *Rhytidiopsis robusta* (Hook.) Broth.; a hand lens will show paraphyllia on the stem of the latter. *Rhytidiopsis* is seldom found below 3000 feet.

X2.5

Rhytidiopsis robusta **Hypnaceae**

ROBUST MOSS

Between 3000 and 4000 feet, the Robust Moss is extremely common, replacing Mountain Fern Moss and Oregon Beaked Moss, which populate the lower elevations. Dr. Conard has informed the author that this was the favorite moss of the University of Washington's late Dr. Frye.

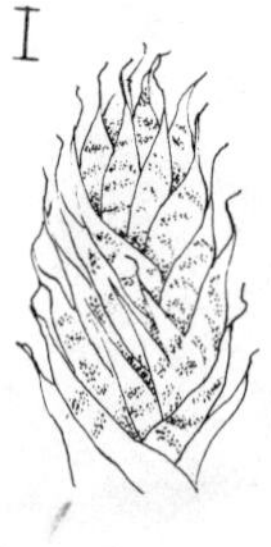

X8

X1.5

Scleropodium obtusifolium **Hypnaceae**

BLUNT-LEAVED MOSS

Common on rocks in river beds. Never continuously under water, but always within easy reach of it.

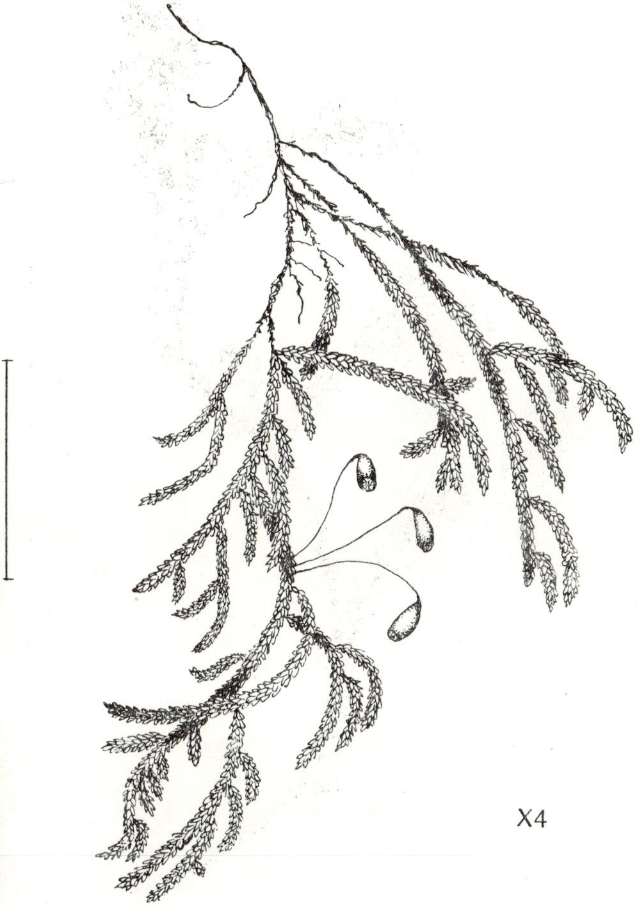

X4

Scouleria aquatica **Grimmiaceae**

SCOULER'S MOSS

Named after Dr. John Scouler, who collected it near Observatory Inlet, British Columbia, in 1829.

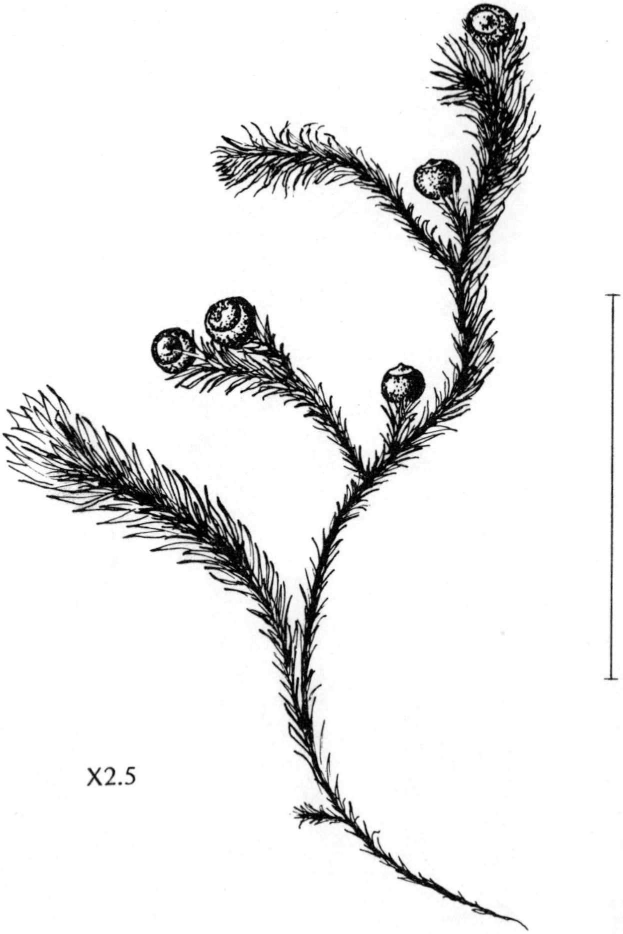

X2.5

Sphagnum girgensohnii **Sphagnaceae**

PEAT MOSS

The peat mosses (Sphagnaceae) are easily recognized but, as in the Bryaceae, the species are difficult for the beginner. Normally associated with swamps, they may be found growing on rocks along the trail in the heavy-rainfall areas on the west side of the Coast Ranges. The most widespread species.

X3/4

Tortula princeps **Pottiaceae**

TWISTED MOSS

The Tortulas are mosses of the drier habitats. Dr. Flowers tells the author that they are particularly common in the dry areas of Utah. The only moss that has a sporophyte as big as that of *T. princeps* is *Bryum sandbergii;* but there the resemblance ends. *Tortula princeps* has a long, tubular peristome that looks almost like a vase, with only the tips of the teeth free. It is common on the stone fireplaces of campgrounds.

X2.5

Ulota obtusiuscula **Orthotricaceae**

OBTUSE ULOTA

Plants in small, dense tufts, yellowish green above and brown below. Leaves tend to curl up when dry.

X6

Ulota phyllantha **Orthotricaceae**

OCEAN ULOTA

The Ulotas are typical of the rain forest. *Ulota* means curled. The author has never found *Ulota phyllantha* outside the reach of the salt spray.

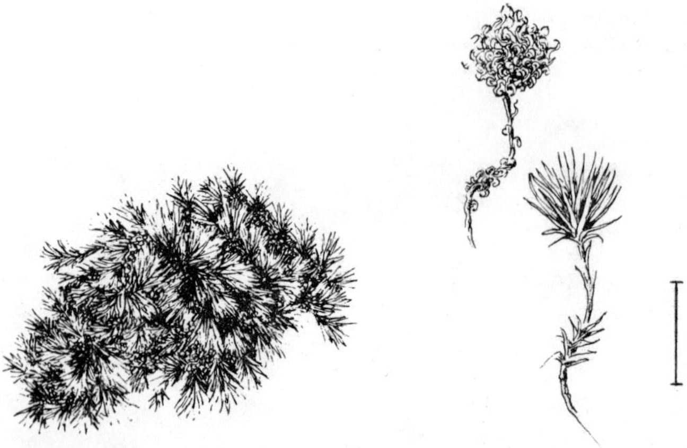

X3

glossary

GLOSSARY

Acrocarpous The condition of bearing the spore capsules on the ends of the stems. (See **pleurocarpous**, below.)

Antheridium That portion of the moss plant that produces sperm or male germ cells.

Archegonium That portion of the moss plant that produces the ovum, or egg.

Calyptra The remains of the archegonium, which forms a cap over the young spore capsule.

Capsule The generally cylindrical vessel that contains the spores.

Chlorophyll The green coloring matter in plants that is the active agent in food making.

Cilia Tiny, hair-like appendages between the inner peristome teeth.

Costa The midrib of the leaf.

Frond In this book, frond refers to a moss that is fern-shaped.

Gamete A germ cell, male or female.

Gametophyte The sexual stage of the moss life cycle: the plant that produces antheridia and archegonia.

Gemmae Little buds from which new plants grow.

Habitat	The environment in which a plant grows naturally.
Hepatic	A relative of the mosses (Liverwort).
Hygroscopic	Sensitive to moisture changes.
Lichen	Plant composed of a fungus and an alga.
Liverwort	Common name for hepatic. (See p. 9.)
Midrib	The 'keel' down the center of a leaf, costa.
Operculum	The cap that covers the peristome teeth on the moss capsule. Shedding of the operculum frees the spores.
Ovum	Egg.
Paraphyllia	Hair- and leaf-like structures that grow among the leaves of some mosses.
Peristome	Teeth that surround the mouth of the spore capsule.
Phenology	A record of the fruiting time of plants.
Pleurocarpous	The configuration of creeping mosses that bear their capsules on setae along the stem. (Contrast with **acrocarpous**, above.)
Protonema	The thread-like plant that germinates from a moss spore; the forerunner of the gametophyte. (Plural: protonemata.)

Protoplasm Living matter.

Rhizoids Root-like structures of mosses.

Rosettes Clusters of leaves that resemble a rose.

Seta The stalk which holds the capsule aloft.

Soredia Tiny pieces of lichen plants that are blown about by the wind, or carried by other agents to form new lichen plants.

Sperm Male germ cell. Generally equipped with a tail or flagellum by which it may swim to the egg.

Spore The asexual (nonsexual) cell from the capsule that starts a new plant (the gametophyte).

Sporophyte The entire part of the plant that produces spores.

Stolon Like the runners on a strawberry plant.

Substrate The immediate material on which a plant is growing.

Tomentum Densely matted, short, woolly hairs.

Vascular In our usage, vascular refers to the flowering plants: having vessels in the roots and stems to conduct water, sap etc.

index

INDEX

INDEX

INDEX

INDEX

INDEX

INDEX

INDEX

NOTES

NOTES

NOTES

NOTES

NOTES